Stefanie S. Hirsch

Elucidating the function of mycorrhizal-induced Kunitz protease inhibitors and characterization of their putative target proteases in *Medicago truncatula*

Elucidating the function of mycorrhizal-induced Kunitz protease inhibitors and characterization of their putative target proteases in *Medicago truncatula*

by
Stefanie S. Hirsch

Dissertation, Karlsruher Institut für Technologie (KIT)
Fakultät für Chemie und Biowissenschaften
Tag der mündlichen Prüfung: 20. Dezember 2013

Impressum

Karlsruher Institut für Technologie (KIT)
KIT Scientific Publishing
Straße am Forum 2
D-76131 Karlsruhe

www.ksp.kit.edu

Print on Demand 2014

ISBN 978-3-7315-0175-6
DOI: 10.5445/KSP/1000038422

Elucidating the function of mycorrhizal-induced Kunitz protease inhibitors and characterization of their putative target proteases in *Medicago truncatula*

Zur Erlangung des akademischen Grades eines
DOKTORS DER NATURWISSENSCHAFTEN

(Dr. rer. nat.)

Fakultät für Chemie und Biowissenschaften
Karlsruher Institut für Technologie (KIT) – Universitätsbereich

genehmigte

DISSERTATION

von

Stefanie S. Hirsch

aus

Villingen-Schwenningen

Dekan:	Prof. Dr. Martin Bastmeyer
Referent:	Prof. Dr. Natalia Requena
Korreferent:	Prof. Dr. Peter Nick
Tag der mündlichen Prüfung:	20.12.2013

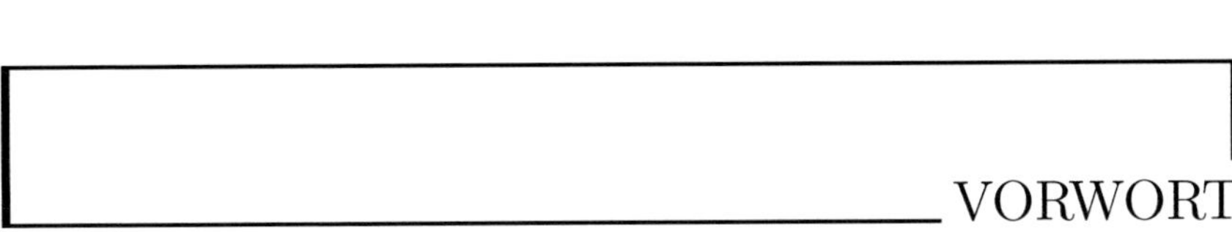

VORWORT

Die Untersuchungen zur vorliegenden Arbeit wurden von Juni 2010 bis Oktober 2013 in der Abteilung Molekulare Phytopathologie des Botanischen Instituts am Karlsruher Institut für Technologie (KIT) unter der Leitung von Prof. Dr. Natalia Requena durchgeführt. Ich habe diese Arbeit selbstständig, unter Beachtung der Satzung des KIT zur Sicherung guter wissenschaftlicher Praxis, angefertigt. Es wurden keine anderen als die angegebenen Quellen und Hilfsmittel benutzt. Die wörtlich oder inhaltlich übernommenen Stellen wurden als solche kenntlich gemacht. Teile dieser Arbeit wurden bereits publiziert[1]. Vorarbeiten zu dieser Arbeit wurden im Rahmen meiner 2009 der Fakultät für Chemie und Biowissenschaften des KIT vorgelegten Diplomarbeit „Elucidating the function of genes encoding secreted proteins induced during early establishment of arbuscular mycorrhiza symbiosis“ geleistet.

[1] **Rech, S.S., Heidt, S., Requena, N. (2013).** A tandem Kunitz protease inhibitor (KPI106)–serine carboxypeptidase (SCP1) controls mycorrhiza establishment and arbuscule development in *Medicago truncatula. Plant J.*, **75**(5), 711–725.

Im Folgenden möchte ich meinen Dank all denjenigen Personen aussprechen, die mich maßgeblich bei der Erstellung dieser Arbeit unterstützt haben. Mein größter Dank gilt Prof. Dr. Natalia Requena, die mir die Möglichkeit geboten hat meine Dissertation über dieses spannende Thema zu verfassen. Natalia, vielen Dank für deine wertvollen Ideen, deine Ratschläge, deine Unterstützung und für all die Dinge, die ich in den letzten Jahren von dir lernen durfte. Herrn Prof. Dr. Peter Nick sei herzlich für die Übernahme des Zweitgutachtens gedankt. Für die finanzielle Unterstützung durch ein Promotionsstipendium danke ich der Landesgraduiertenförderung Baden-Württemberg. Besonders bedanken möchte ich mich bei Sven Heidt, der im Rahmen seiner Bachelorarbeit die Proteinmodelle erstellt, die Modellierung der Protein–Protein Interaktion durchgeführt und mich bei der Klonierung einiger Plasmide unterstützt hat. Vielen Dank für deine Mitarbeit an diesem Projekt. Bei Herrn Prof. Dr. Renier van der Hoorn möchte ich mich für seine hilfreichen Ratschläge zu diesem Projekt bedanken. Ich danke allen ehemaligen und aktuellen Mitarbeiterinnen und Mitarbeitern des Myclabs für die wunderbare gemeinsame Zeit im Labor und für die aufschlussreichen Diskussionen in unseren wöchentlichen Seminaren, die auch die Arbeitsgruppen von Prof. Dr. Reinhard Fischer und Prof. Dr. Jörg Kämper miteingeschlossen haben. Zum Schluss möchte ich mich bei meiner Familie für ihre Liebe und ihre volle Unterstützung in allen Lebenslagen bedanken. Danke auch dafür, dass ihr immer an mich geglaubt habt. Aus tiefstem Herzen danken möchte ich meinem Mann für seine Liebe, seine Geduld, sein Verständnis und für seine Motivation, ohne die diese Arbeit nicht möglich gewesen wäre.

Karlsruhe, Februar 2014 *Stefanie S. Hirsch*

ABSTRACT

The majority of terrestrial plants live in symbiosis with arbuscular mycorrhizal fungi. Those fungi form highly branched intraradical structures, called arbuscules, in order to establish a bidirectional nutrient exchange with their host plant. In this study, a mycorrhizal-induced gene family of *Medicago truncatula* encoding putative Kunitz protease inhibitors (KPI), was functionally characterized. Those types of protease inhibitors were reported to be involved in various interactions of plants with other organisms, including defense responses against herbivory insects and pathogenic microbes. Interaction tests in yeast revealed that the putative target proteases of the mycorrhizal-induced KPIs are among a clan of serine carboxypeptidases (SCP) that are also exclusively expressed during mycorrhizal symbiosis. This suggested the possibility for tandem interactions so that each family member might have a defined counterpart. *In silico* modeling of the interaction between KPI106 and the serine carboxypeptidase SCP1 revealed that KPI106 is likely a canonical inhibitor, as a distinct loop containing the predicted reactive site residues P_1 Lys^{173}, $P_{1'}$

Phe^{174} and P$_{2'}$ Glu^{175}, enters the active pocket of the protease SCP1 in a substrate-like manner. *In vitro* mutational analysis confirmed the importance of Lys^{173} in mediating strength and specificity of the interaction, and showed that Phe^{174} and Glu^{175} are even necessary for the interaction with SCP1. Heterologous expression of SCP1, KPI106 and the homologous inhibitor KPI104 in tobacco epidermal cells revealed that all proteins are likely secreted. Furthermore, deregulation of the spatio-temporal expression of KPI106 and KPI104 led to an impaired symbiotic development including malformed arbuscules and septated fungal hyphae; a phenotype that was also observed by RNAi-mediated silencing of the SCP family. Based on the similarity of those phenotypes, it is suggested that the KPIs and SCPs might work together on the same process. A subsequent genetic analysis of the KPI overexpression and SCP-silenced phenotypes revealed that those can be categorized into a group of mutant plants, in which arbuscule development is impaired in an early branching state. It is thus proposed that the KPIs would regulate the spatial and temporal activity of the SCPs. In turn, the SCPs would produce a bioactive peptide signal, which is involved in the control of mycorrhizal establishment and arbuscule development within the root cortex of *M. truncatula*.

ZUSAMMENFASSUNG

Die meisten Landpflanzen leben in einer Symbiose mit arbuskulären Mykorrhizapilzen. Diese Pilze bilden hochverzweigte Strukturen im Inneren der Pflanzenwurzel aus, die sogenannten Arbuskel, um einen gegenseitigen Nährstoffaustausch zu ermöglichen. In dieser Arbeit wurde eine Proteinfamilie von Kunitz-Proteaseinhibitoren (KPI) funktionell charakterisiert, deren Mitglieder spezifisch in *Medicago truncatula* während der arbuskulären Mykorrhizasymbiose exprimiert werden. Kunitz-Proteaseinhibitoren wurden im Zusammenhang mit einer Vielzahl von biologischen Prozessen in der Pflanze beschrieben, unter anderem auch in der Abwehr gegen herbivore Insekten und pathogene Mikroorganismen. Durch Interaktionstests in Hefe konnte gezeigt werden, dass mehrere Mitglieder einer Familie von mykorrhiza-induzierten Serin-Carboxypeptidasen (SCP) zu den putativen Zielproteasen dieser Kunitz-Inhibitoren gehören. Dies deutet darauf hin, dass eventuelle Tandem-Interaktionen zwischen den Mitgliedern dieser beiden Proteinfamilien stattfinden. Eine *in silico* Modellierung der Protein-Protein Interaktion zwischen KPI106 und der Serin-Carboxypeptidase

SCP1 ergab, dass KPI106 wahrscheinlich ein kanonischer Proteaseninhibitor ist, da eine charakteristische Peptidschleife, welche die vorhergesagten reaktiven Aminosäuren P_1 Lys^{173}, $P_{1'}$ Phe^{174} und $P_{2'}$ Glu^{175} enthält, substratähnlich in das aktive Zentrum der Protease SCP1 eindringt. Durch *in vitro* Mutationsanalysen konnte gezeigt werden, dass Lys^{173} von KPI106 einen Einfluss auf die Spezifität und die Interaktionsaffinität mit SCP1 hat. Des Weiteren sind Phe^{174} und Glu^{175} sogar notwendig um eine Interaktion von KPI106 und SCP1 zu ermöglichen. Die heterologe Expression von KPI106, des homologen Inhibitors KPI104 und SCP1 in epidermalen Tabakzellen zeigte, dass alle Proteine sehr wahrscheinlich sekretiert werden. *M. truncatula* Wurzeln, in denen die Genexpression von KPI106 und KPI104 dereguliert wurde, wiesen einen Phänotyp in der Mykorrhizasymbiose auf, welcher mit einer reduzierten Arbuskelhäufigkeit und dem Auftreten von septierten intraradikalen Hyphen einhergeht. Ein sehr ähnlicher Phänotyp konnte in Wurzeln beobachtet werden, in denen die Menge an SCP Proteinen durch RNAi-vermittelte Genunterdrückung reduziert wurde. Die genetische Analyse des KPI-Überexpressions- und SCP-RNAi Phänotyps ergab, dass dieser einer Gruppe von Arbuskelmutanten zugeordnet werden kann, deren Arbuskelentwicklung in einem frühen Verzweigungsstadium blockiert ist. Daher wurde die Hypothese aufgestellt, dass die mykorrhiza-induzierten Kunitz-Proteaseinhibitoren die räumliche und zeitliche Aktivität der SCP Proteine regulieren. Deren Aufgabe wiederum wäre es ein bioaktives Signal zu prozessieren, welches wichtig für die Etablierung der Mykorrhizasymbiose, im Besonderen für die Arbuskelentwicklung im Wurzelcortex von *M. truncatula* ist.

CONTENTS

LIST OF FIGURES

GLOSSARY

α	Anti
aa	Amino acid
AM	Arbuscular mycorrhizal
CO	Chitooligosaccharides
Conc.	Concentration
CP	Cysteine protease
DBD	DNA-binding domain
DEPC	Diethylpyrocarbonate
DFCI	Dana Farber Cancer Institute
DMI	Does not make infections
DMSO	Dimethyl sulfoxide
dpi	Days post infection
GEA	Gene expression atlas
GST	Glutathion-S-transferase
GST	Glutathione S-transferase
GUS	β-glucuronidase
His	Histidine
ID	Identity
IMGAG	International *Medicago* Gene Annotation Group
KPI	Kunitz protease inhibitor
LCO	Lipo-chitooligosaccharides
LRR	Leucine-rich repeat
LysM	Lysine motif
Mbp	Mega base pairs
MST	Monosaccharide transporter
MtGI	*Medicago truncatula* Gene Index
NCBI	National Center for Biotechnology Information

Nod	Nodulation
nr	Non-redundant
p35S	Cauliflower mosaic virus 35S promotor
PAM	Periarbuscular membrane
PAS	Peri-arbuscular space
PDB	Protein Data Bank
PMSF	Phenylmethylsulfonyl fluoride
PPA	Prepenetration apparatus
PT4	Phosphate transporter 4
pUbi	Ubiquitin promoter
RCSB	Research Collaboratory for Structural Bioinformatics
RT	Room temperature
SbtM1	Subtilase M1
SCP	Serine carboxypeptidase
SD	Synthetic defined
STI	Soybean trypsin inhibitor
TAD	Transactivation domain
TC	Tentative consensus
WGA	Wheat germ agglutinin

CHAPTER 1

INTRODUCTION

1.1 Arbuscular mycorrhizal symbiosis

The arbuscular mycorrhizal (AM) symbiosis is a beneficial association between the majority of plants and fungi of the Glomeromycota phylum (Schüßler *et al.*, 2001). The main goal of this endosymbiosis is a bidirectional nutrient exchange (Smith & Read, 2008). Therefore, the fungal hyphae need to penetrate the plant root through the rhizodermal layer and grow deeper into the root interior up to the inner cortical cells. Thereupon, fungal hyphae undergo extensive branching and form tree-like structures—the arbuscules. Those intraradical fungal structures were name giving for the symbiosis and are considered as the symbiotic interface where most of the nutrient exchange takes place.

1.1.1 The mutual benefits

AM fungi are obligate biotrophs and rely on a plant host for the supply of carbon and to complete their life cycle (Requena *et al.*, 2007). Thus, the plant delivers up to 20 percent of its photosynthetically fixed carbon (Bago *et al.*, 2000) to the fungal partner. In return, the fungus—due to the extensive extraradical mycelium (Miller *et al.*, 1995)—provides its host plant mainly with phosphate and nitrogen, as well as with water from out the soil (Finlay, 2008). In addition, plants colonized with AM fungi were reported to be less susceptible against phytopathogens (Whipps, 2004; Liu *et al.*, 2007) and various abiotic stress factors (Giri *et al.*, 2007).

To manage the bidirectional nutrient exchange, the arbuscules are surrounded by a modified plant-derived membrane, the peri-arbuscular membrane (PAM; Figure 1.5; Bonfante & Perotto, 1995), which carries an unique assembly of symbiotic phosphate and nitrogen transporters (Harrison, 2012; Harrison *et al.*, 2002; Guether *et al.*, 2009b; Kobae *et al.*, 2010). Recently, Helber *et al.* (2011) identified the fungal monosaccharide transporter 2 (MST2), that is also highly abundant in arbuscules. Furthermore, they showed that the expression of *MST2* is correlated with that of the symbiotic plant phosphate transporter 4 (*PT4*; Harrison *et al.*, 2002), and that silencing of *MST2* led to an impaired mycorrhiza formation with a reduced *PT4* expression. Consequently, Helber *et al.* (2011) propose that the fungal sugar uptake from the plant is correlated with the supply of phosphate.

1.1.2 The long story of mycorrhizal symbiosis

The scientific term mycorrhiza was introduced by Bernhardt Frank in 1885. He observed that the end branches of the root systems of trees were completely enclosed in a continuous cover of fungal hyphae, which he referred to as fungus-root or mycorrhiza (original publication: Frank,

1885; english translation: Frank, 2005). Based on sequence data and fossil records, the origin of this symbiosis can be tracked back over 450 million years to the age of the Ordovician (Remy *et al.*, 1994, Redecker *et al.*, 2000). During this geologic period, the first primitive land plants appeared (for example: Rubinstein *et al.*, 2010). It is thus postulated that arbuscular mycorrhizae supported archaic plants in the colonization of land by providing nutrients in the absence of a complex vascular root system (Simon *et al.*, 1993; Wang *et al.*, 2010).

To date, with the exception of a few plant species that do not engage in mycorrhizal symbiosis (Wang & Qiu, 2006), the mycorrhizal state is still default for plants growing in most field conditions (Smith & Smith, 2011). This indicates an ancient co-evolution and that the symbiosis had a selective advantage for both partners (Smith & Read, 2008). As a consequence of this broad distribution, the mycorrhizal symbiosis is attributed to have an influence on the global carbon cycle (Harrison, 2012). Furthermore, with regard to a sustainable agriculture, the possible role of AM symbiosis in plant phosphorous nutrition and thus contributing to plant growth and productivity, raised the agronomic interest (Sawers *et al.*, 2007; Smith & Smith, 2011).

1.2 AM symbiosis in the model legume *Medicago truncatula*

Medicago truncatula, also referred to as barrel medic, is endemic to the Mediterranean region (Lesins & Lesins, 1979). It belongs to the legume family that also includes members with major impact on world agriculture, such as soybean, chickpea, and alfalfa. However, due to the size and ploidy level of their genomes, it is difficult to genetically characterize those crops. In contrast, *M. truncatula* offers several advantages as it contains a simple diploid genome with eight sets of chromosome pairs, containing 500 to 550 Mbp in total. It is furthermore self-fertilizing with a short seed-to-seed generation time and due to its ease for genetic transformation, various molecular and genomic tools are available. Moreover, unlike the existing plant model *Arabidopsis thaliana*, legumes can establish endomycorrhiza and in addition undergo a symbiotic relationship with nitrogen fixing bacteria to form root nodule symbiosis (Schüßler, 2009). Since both symbioses share common molecular features (Kistner *et al.*, 2005), it is important to study both plant-microbe interactions within the same plant; making *M. truncatula* an ideal model for the legume family (Cook, 1999).

The experiments of this study were also performed in *M. truncatula,* and furthermore, the arbuscular mycorrhizal fungus *Rhizophagus irregularis* was used as symbiont to study the effects on mycorrhizal symbiosis. Figure 1.1 presents an overview of arbuscular mycorrhiza formation within the root cortex of *M. truncatula* in association with *R. irregularis*, highlighting the different symbiotic stages and life cycle of arbuscular mycorrhizal fungi. In the following subsections the current state of the art of molecular and cellular aspects of AM development will be introduced. The main focus shall be put on arbuscule development that is of particular interest to integrate the results of this work.

1.2.1 Molecular communication

The first step in mycorrhizal symbiosis is the reciprocal recognition of plant and fungus that is carried out by the exchange of signaling molecules. Plant roots permanently produce and secrete a mixture of substances into the rhizosphere, the root exudates (Bais *et al.*, 2006). Among those components, it is mainly strigolactone, a carotenoid-derived phytohormone, that stimulates germination and subsequent hyphal branching of AM fungal spores in the vicinity of roots (Akiyama *et al.*, 2005; Besserer *et al.*, 2006;

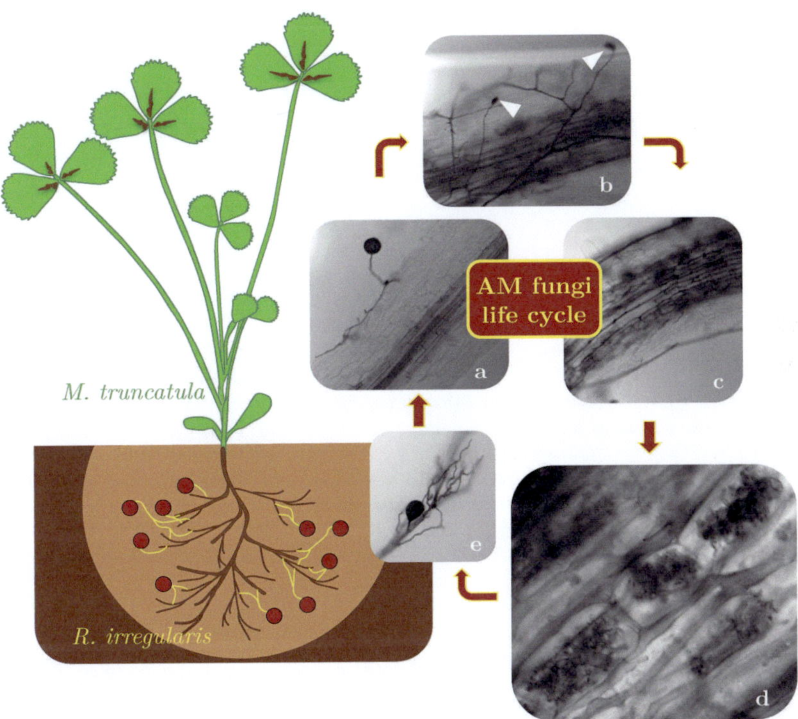

Figure 1.1: AM formation in the root cortex of *M. truncatula*. Left: Schematic diagram of *M. truncatula* roots colonized by the soil-inhabiting AM fungus *R. irregularis* (red: spores; yellow: fungal hyphae). **Right:** Life cycle of AM fungi, using the example of *R. irregularis*. **(a)** Germinated fungal spore making first hyphal contact to the host root. **(b)** Hyphal branching and formation of hyphopodia (white arrow) on the root surface. **(c)** Intraradical hyphae growing through the *M. truncatula* root cortex. **(d)** Differentiation of fungal hyphae into highly branched arbuscules within the inner cortical cells. **(e)** Extraradical mycelium of *R. irregularis* that develops new spores for subsequent root colonization.

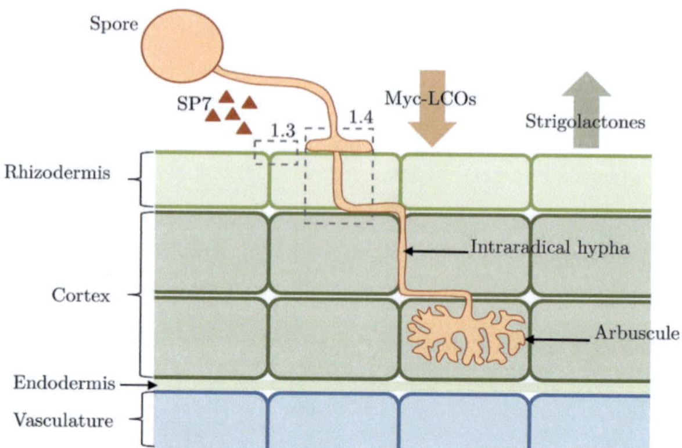

Figure 1.2: Establishment of AM symbiosis, schematic. Root-derived strigolactones induce spore germination in AM fungi that secrete Myc-LCOs to communicate with the plant. Additionally, AM fungi secrete the effector SP7 to promote colonization. Thereupon, fungal hyphae penetrate the root surface *via* hyphopodium, traverse the rhizodermal cell layer intracellularly. They grow intercellularly through the apoplast towards the inner cortex, where they differentiate into the highly branched arbuscules. The dashed rectangles indicate two cutouts that are presented in Figure 1.3 and Figure 1.4 in more detail.

Kretzschmar *et al.*, 2012). Furthermore, two hydroxy fatty acids were reported to induce hyphal branching in the AM fungus *Gigaspora gigantea* (Nagahashi & Douds Jr., 2011).

On the fungal side, upon germination, AM fungi secrete diffusible molecules that lead to the activation of a symbiotic plant program in order to prepare the rhizodermal cells for the entering symbiont (Figure 1.2; Bonfante & Requena, 2011). The chemical nature of these molecules is a combination of N-acetylglucosamine oligosaccharides (Maillet *et al.*, 2011; Genre *et al.*, 2013), whereas especially the lipo-chitooligosaccharides (LCOs) stimulate AM symbiosis (Maillet *et al.*, 2011). The LCOs derived from AM fungi (referred to as Myc factors) are structurally similar to the secreted Nod factors of nitrogen fixing rhizobia (Lerouge *et al.*, 1990),

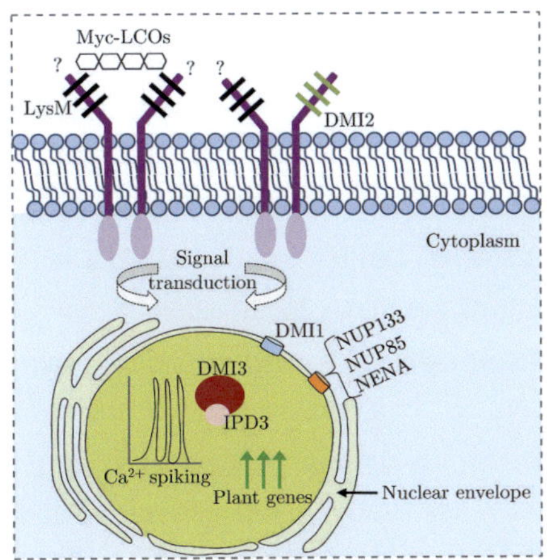

Figure 1.3: The common signaling pathway, simplified diagram. The Myc-LCOs are probably perceived *via* a LysM receptor-like kinase (or a dimer) that might build together with the LRR receptor kinase DMI2 a receptor complex within the plasma membrane. Intracellular signal transduction involves the nuclear pore associated NUP133, NUP85, and NENA, as well as the nuclear membrane-localized cation channel DMI1. A following nuclear calcium spiking activates the calcium-calmodulin dependent kinase DMI3 that interacts with IPD3. The common signaling pathway culminates in the induction of symbiosis-specific plant genes. Diagram modified from Oldroyd (2013).

which are required to establish root nodule symbiosis (Denarie *et al.*, 1996; Gough & Cullimore, 2011). Perception of the respective LCOs by the plant is transduced *via* a common signaling pathway, which is shared by both symbioses and is necessary to promote the further colonization process (Catoira *et al.*, 2000; Kistner *et al.*, 2005; Gutjahr & Parniske, 2013).

1.2.2 Signal transduction

The common signaling pathway includes the membrane-bound leucine-rich repeat (LRR) receptor kinase DMI2 (does not make infections; Endre *et al.*, 2002; Stracke *et al.*, 2002), three nuclear pore associated proteins

LjNUP85 (Saito *et al.*, 2007), LjNUP133 (Kanamori *et al.*, 2006), LjNENA (Groth *et al.*, 2010) and additionally DMI1, a nuclear membrane-localized cation channel (Ané *et al.*, 2004; Charpentier *et al.*, 2013). Those proteins are essential for the induction of a nuclear calcium spiking as a second messenger that is characteristic for either rhizobia or AM fungi (Kosuta *et al.*, 2008; Chabaud *et al.*, 2011; Genre *et al.*, 2013). Further downstream of the calcium spiking, the nuclear-localized calcium-calmodulin dependent kinase DMI3 (Lévy *et al.*, 2004; Mitra *et al.*, 2004) is activated and interacts with IPD3, a protein with a yet unknown function (Messinese *et al.*, 2007). Finally, the common signaling pathway leads to the transcriptional activation of symbiosis-associated plant genes (Kosuta *et al.*, 2003; Kuhn *et al.*, 2010), required for colonization of the outer cell layers. Further plant responses are marked by starch accumulation in roots (Gutjahr *et al.*, 2009) and lateral root formation (Oláh *et al.*, 2005). A schematic overview of the common signaling pathway is presented in Figure 1.3.

1.2.3 Friend or foe?

Up until now, the question about how the Myc-LCOs are perceived by the plant remains elusive. However, DMI2 is proposed to be associated with a yet-unknown Myc-LCO receptor (Figure 1.3; Oldroyd, 2013), as *dmi2* mutants are implicated in establishment of mycorrhizal and root nodule symbiosis (Catoira *et al.*, 2000). It is furthermore assumed that this yet unknown plant receptor for Myc-LCO perception is part of a family of lysine motif (LysM) receptor-like kinases (Figure 1.3), because the responses of rhizobia-derived Nod factors rely on two LysM receptor-like kinases, named NFP and LYK3 in *M. truncatula* (Broghammer *et al.*, 2012). In support of this, Op den Camp *et al.* (2011) showed that silencing of a LysM receptor-like kinase in *Parasponia andersonii* abolished nodulation as well as AM formation, and Fliegmann *et al.* (2013) revealed that the

LysM receptor-like kinase LYR3 exhibits a high binding affinity for both, Myc-LCOs and rhizobia-LCOs (as does NFP; Maillet *et al.*, 2011), but not for general chitooligosaccharides (COs). COs are parts of the fungal cell wall polymer chitin and are known to elicit a MAMP-triggered (microbe-associated molecular patterns) plant defense response (Boller, 1995; Boller & Felix, 2009). Thereby, the LysM receptor-like kinase AtCERK of *A. thaliana* is essential for the recognition of chitin (Miya *et al.*, 2007; Wan *et al.*, 2008). Ultimately, N-acetylglucosamine oligosaccharides and thus the secreted LCOs of AM fungi and rhizobia are important signatures for plants to fine-tune their response upon microbial penetration. Thus, in case of symbiotic microbes, the plant promotes the association, whereas in case of pathogenic invaders the lack of LCOs might elicit a defense response. It is furthermore proposed that the LysM receptors for sensing the secreted LCOs of symbiotic microbes have evolved from LysM receptors that activate plant immunity, and the investigation of their biochemical potential and the related intracellular signaling is part of the ongoing research (Gust *et al.*, 2012).

1.2.4 Intraradical accommodation of the fungus

The above mentioned common signaling pathway generates molecular signals necessary for the plant to engage in symbiosis and to promote the colonization of the outer cell layers. On the fungal side, for penetration of the rhizodermis, AM fungal hyphae need to form a hyphopodium. Hyphopodial differentiation is thereby guided by root cell wall components (Nagahashi & Douds Jr., 1997), in particular cutin monomers (Gobbato *et al.*, 2012; Wang *et al.*, 2012). Mutant plants in which either of the gens *RAM2*, encoding a glycerol-3-phosphate acyl transferase involved in the production of cutin biosynthesis, or *RAM1*, encoding a transcription factor that regulates *RAM2*, is affected, showed a reduced hyphopodia

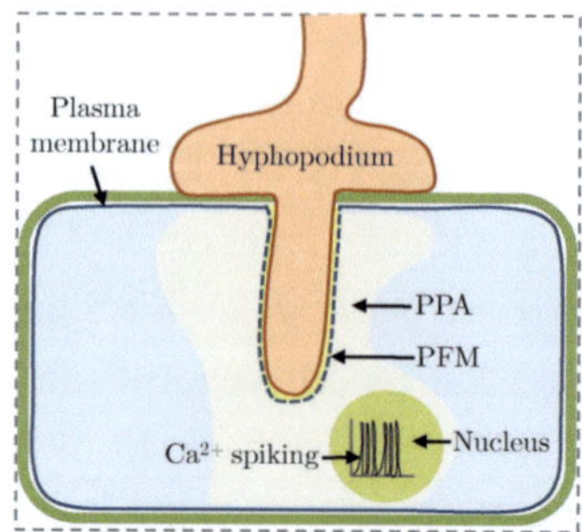

Figure 1.4: Formation of the prepenetration apparatus, simplified diagram. In anticipation of fungal penetration, the rhizodermal cells beneath the hyphopodium differentiate. They form a tubular, transvascuolar structure, the so-called prepenetration apparatus (PPA), which guides the fungal hypha through the cell. Intracellular fungal hyphae are surrounded by plant-derived membrane, referred to as peri-fungal membrane (PFM). Activated rhizodermal cells are marked by a switch of the nuclear calcium spiking from a low to a high frequency. However, it remains elusive whether PPA formation occurs before or is the result of this change of frequency. Diagram modified from Harrison (2012).

formation (Gobbato *et al.*, 2012; Wang *et al.*, 2012). In parallel to the hyphopodium formation, the underlying rhizodermal cells undergo re-differentiation processes in anticipation of the fungal penetration. Thereby, a tunnel-like transvascuolar structure, called the prepenetration apparatus (PPA; Figure 1.4) is formed, which predetermines the path of the fungal hyphae through the rhizodermal cell (Genre *et al.*, 2005). So far, it is not known what signal leads to PPA formation. However, it seems that PPA induction is a very early event, as PPA formation is abolished in the common signaling mutants *dmi2* and *dmi3* (Genre *et al.*, 2005). In line with this, Takeda *et al.* (2012) showed that a constitutive active version of DMI3 is sufficient to trigger a re-differentiation of cortical cells that resembles PPA formation: In turn, this implicates that DMI3 would induce PPA formation. In contrast to those results, Sieberer *et al.* (2012) observed a switch from a low-frequency calcium spiking in "activated" rhizodermal cells to a high-frequency calcium spiking in PPA containing cells. Therefore, the authors postulate that this change in spiking nature could indicate the irreversible commitment of the respective plant cell for intracellular infection. In consequence, this would mean that activation of DMI3 would occur after PPA formation; and so far, it remains elusive how the signals of the common signaling pathway are integrated in relation to the cellular events during fungal penetration (Gutjahr & Parniske, 2013).

For the proceeding intracellular colonization, the gene *VAPYRIN* is essential, as fungal penetration is abolished in the respective mutants. However, in some cases, fungal hyphae could pass the rhizodermis but arbuscule formation is aborted (Reddy *et al.*, 2007; Pumplin *et al.*, 2010). It is thus hypothesized that VAPYRIN is involved in protein interactions during cellular reorganization to support fungal development (Pumplin *et al.*, 2010; Feddermann & Reinhardt, 2011). A similar phenotype was observed when silencing the mycorrhizal-induced membrane-steroid binding protein 1 of *M. truncatula*. In this case, colonized roots revealed defects in appressoria formation accompanied by a reduced number of arbuscules with a distorted morphology. The authors suggest that induction of membrane-steroid binding protein 1 is required to alter the sterol metabolism which would be necessary for an invagination of the plasma membrane during the intracellular accommodation of the fungus (Kuhn *et al.*, 2010). Moreover, very recent findings indicate that also an E3 ubiquitin ligase (Takeda *et al.*, 2013) and two transcription factors, named NSP1 (Delaux *et al.*, 2013; Takeda *et al.*, 2013) and NSP2 (Liu *et al.*, 2011; Maillet *et al.*, 2011) are required for the proceeding of fungal colonization within the root cortex.

A feature provoked by AM fungi to facilitate their intraradical accommodation is indicated by the recent finding of Kloppholz *et al.* (2011), who identified the first AM fungal effector protein SP7 (secreted protein 7; Figure 1.2). When expressed *in planta*, SP7 localized to the plant nucleus and interacted with the ethylene response factor 19 (ERF19). ERF19 is a defense-related transcription factor and overexpression of ERF19 impaired mycorrhizal root colonization. Thus, it was proposed that, in analogy to the effectors of microbial pathogens that are secreted to circumvent or suppress a plant defense response and to promote compatibility (Stergiopoulos & de Wit, 2009), SP7 would promote mycorrhizal colonization by suppression of ERF19-related defense responses (Kloppholz *et al.*, 2011). This is in line with ethylene being a negative regulator of AM colonization (Foo *et al.*, 2013; Gutjahr & Parniske, 2013; Mukherjee & Ané, 2010; Penmetsa *et al.*, 2008).

1.2.5 Arbuscule development

After passing the root surface and growing into the deeper root layers, the fungal hyphae enter the inner cortical cells. As observed for the initial rhizodermal penetration, the fungal hyphae are guided by a PPA-like structure towards the center of the inner cortical cells (Genre *et al.*, 2008). Thereupon, they switch from polar growth to dichotomous hyperbranching in order to form the arbuscule. At the end, a mature arbuscule almost fills out the entire cellular interior, which in turn, imposes drastic reorganization processes on the inner cortical cells (Cox & Sanders, 1974; Pumplin & Harrison, 2009).

1.2.5.1 Subcellular reorganization of the inner cortical cells

To enable intracellular development of the hyperbranching fungal hypha, the organelles of the inner cortical cells have to be rearranged (overview in Pumplin & Harrison, 2009). Furthermore, the plant plasma membrane invaginated by the developing arbuscule has to be enlarged in surface in order to envelope the growing fungal structure (Cox & Sanders, 1974; Scannerini & Bonfante-Fasolo, 1983). This is mirrored by the unique transcriptional profile of arbusculated inner cortical cells, of which many of the differentially regulated genes are related to transport processes and lipid metabolism (Gomez *et al.*, 2009; Hogekamp *et al.*, 2011; Gaude *et al.*, 2012). Thereby, among others, the transition of the arbuscule-surrounding plant plasma membrane to the PAM is marked by the assembly of transporter proteins for the symbiotic nutrient exchange, or to export substrates into the peri-arbuscular space (PAS; Figure 1.5; Harrison, 2012). Those transporters include the two ABC half-size transporters STR and STR2 (Zhang *et al.*, 2010; Gutjahr *et al.*, 2012) and the extensively characterized *M. truncatula* PT4 (Harrison *et al.*, 2002), which is essential for the symbiotic phosphate uptake by the plant (Javot *et al.*, 2007). The exclusive targeting of the transporter PT4 to the PAM is mediated by the precise timing of its expression coupled with a transient remodeling of vesicle trafficking in order to favor secretion to the PAM over localization to the plant plasma membrane (Pumplin *et al.*, 2012). In line with this, Genre *et al.* (2012) and Ivanov *et al.* (2012) elucidated that exocytotic-vesicle-associated membrane proteins, VAMPs, are necessary for PAM formation, and Lota *et al.* (2013) showed that silencing of a member of the SNARE family (proteins involved in targeting of transport vesicles and fusion with the destination membrane; overview in Südhof & Rothman, 2009) leads to a distorted arbuscule phenotype in the legume *Lotus japonicus*.

1.2.5.2 Arbuscule lifetime

The arbuscule is a transient structure, and after a period of maturity arbuscules collapse and die. During this process, the inner cortical cells stay intact and can host successive generations of arbuscules (Cox & Sanders, 1974; Bonfante-Fasolo, 1984; Javot *et al.*, 2007). The average lifetime of arbuscules is first of all depending on the plant host. Whereas in *M. truncatula*, the average arbuscule life time is about eight days (Alexander *et al.*, 1989; Javot *et al.*, 2007), they only last between two to three days in rice (Kobae *et al.*, 2010). Secondly, the efficiency of the symbiotic nutrient transfer seems to prolong arbuscule lifetime. The first and most prominent marker gene of mature and active arbuscules is *PT4* (Harrison *et al.*, 2002). The arbuscules in colonized *pt4* mutant plants are malformed and contain septa, indicating their premature collapse as a consequence of the prohibited symbiotic phosphate uptake (Javot *et al.*, 2007). In addition, an insufficient supply of carbon to the fungus may be associated with a premature arbuscule collapse (Gutjahr & Parniske, 2013). In support of this, silencing of the *M. truncatula* symbiosis-induced sucrose synthase 1 (Baier *et al.*, 2010), or the previously mentioned fungal MST2 (Helber *et al.*, 2011) revealed a malformed arbuscule phenotype, accompanied by a significant reduction of *PT4* expression (Baier *et al.*, 2010; Helber *et al.*, 2011). In contrast, there are mutant plants that exhibit an arbuscule phenotype but the expression of *PT4* is not altered (Zhang *et al.*, 2010; Gutjahr *et al.*, 2012; Groth *et al.*, 2013). This indicates that *PT4* and thus symbiotic nutrient transfer is necessary but not sufficient for arbuscule formation and argues for additional, *PT4* independent, mechanisms that are required for arbuscule development (Gutjahr & Parniske, 2013).

1.2.5.3 Current model of arbuscule development

In contrast to rhizodermal penetration and the early colonization stages, arbuscule development is not directly activated *via* the common signaling pathway (summarized in Harrison (2012); Gutjahr & Parniske (2013)). In particular, with regard to the common signaling mutants *dmi2, dmi1, nup133, nup185* and *nena* (Wegel *et al.*, 1998; Demchenko *et al.*, 2004; Kistner *et al.*, 2005; Groth *et al.*, 2010; see Figure 1.3), arbuscule development in those mutants is not affected once the fungus has successfully overcome colonization of the rhizodermal and outer cortical layers. This suggests that the common signaling pathway is mainly restricted to control the colonization of the outer root parts. In support of this, a gain-of-function version of the common signaling protein DMI3 is able to suppress the loss-of-function mutations of the upstream common signaling genes (Hayashi *et al.*, 2010). Furthermore, the DMI3 gain-of-function version is able to induce a pre-arbuscular cortical cell differentiation (Takeda *et al.*, 2012). This indicates that the upstream common signaling genes are solely required to activate DMI3 and are not involved in the induction of arbuscule-related cell differentiation (Hayashi *et al.*, 2010; Takeda *et al.*, 2012).

In order to categorize the present arbuscule mutants, Gutjahr & Parniske (2013) propose a model to divide arbuscule development into five genetically separable stages; a detailed scheme is presented in Figure 1.5. The first two stages are thereby characterized by the formation of a PPA-like structure in the inner cortical cell, followed by fungal penetration (Genre *et al.*, 2008). Accordingly, the first stage is genetically characterized by the expression of the marker genes *DMI3* and *IPD3*, as the respective mutant plants showed a complete loss in arbuscule formation (Demchenko *et al.*, 2004; Kistner *et al.*, 2005; Yano *et al.*, 2008). The subsequent third stage is named "birdsfoot", as the initial arbuscule branches, which are characteristic for this stage, resemble the structure of a birdsfoot. The expression of the marker gene *VAPYRIN* is essential for stage 2 and 3, as arbuscule development in *vapyrin* mutants is restricted to hyphal knobs (Reddy *et al.*, 2007; Pumplin *et al.*, 2010). Continuous dichotomous hyphal branching in stage 3 leads over to stage 4, the mature arbuscule. Marker genes of stage 4 are the previously mentioned *STR* and *STR2* (Zhang *et al.*, 2010; Gutjahr *et al.*, 2012), as well as the affected genes in the mutants *SL0154* and *SL0181* (Groth *et al.*, 2013), because mutation of either of those genes led to a developmental arrest of arbuscules in the birdsfoot stage. Furthermore, stage 4 requires the expression of *RAM2*, as *ram2* mutants showed—in addition to the already mentioned hyphopodia

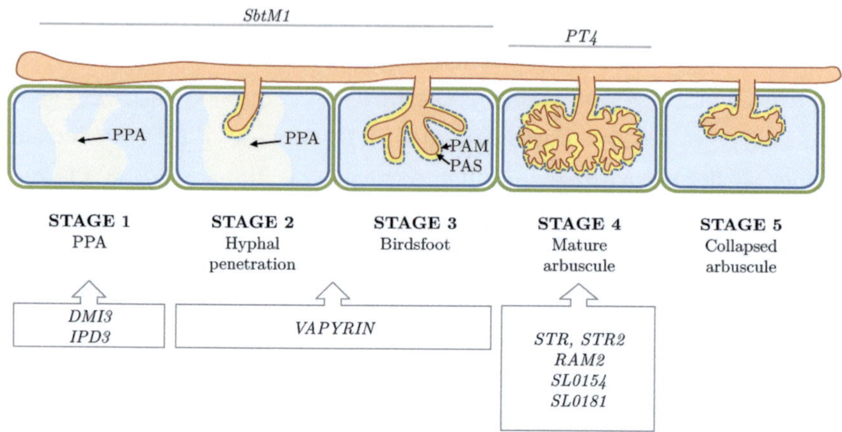

Figure 1.5: Different stages of arbuscule development (Gutjahr & Parniske, 2013). This simplified model is based on Gutjahr & Parniske (2013), and illustrates the five genetically separable stages of arbuscule development. PPA formation in stage 1 is dependent on *DMI3* and *IPD3* expression, whereas *VAPYRIN* is required for stage 2 (hyphal penetration) and stage 3 (birdsfoot) in which the arbuscule branching is initiated. For development of the mature arbuscule, marked by stage 4, the genes *STR, STR2, RAM2* as well as the genes affected in the mutants *SL0154* and *SL0181* are necessary. Arbuscule development is further overlaid by two distinct waves of gene induction. The first wave, marked by *SbtM1*, is probably induced by the common symbiosis pathway, whereas the second wave is most likely independent of the common symbiosis pathway and is characterized by *PT4* expression which delays arbuscule senescence (stage 5). Blue dashed line: PAM; In yellow: PAS.

defect—occasional colonization events with an impaired arbuscule formation (Wang *et al.*, 2012). Ultimately, *PT4* expression in stage 4 delays arbuscule senescense, referred to as stage 5. In *pt4* mutants, arbuscule senescence is induced prior maturity (Javot *et al.*, 2007).

In addition, Gutjahr & Parniske (2013) suggest that arbuscule development is overlaid with two independent waves of gene induction. Among others, one marker gene of the first wave of gene induction is the subtilase *SbtM1* (Takeda *et al.*, 2007; Takeda *et al.*, 2009; Takeda *et al.*, 2012). Silencing of *SbtM1* led to an impaired arbuscule formation and a lower fungal colonization (Takeda *et al.*, 2009). Because *SbtM1* is induced in the DMI3 gain-of-function mutant (Takeda *et al.*, 2012), Gutjahr & Parniske (2013) propose that this first wave of gene induction is directly or indirectly controlled by the common signaling pathway. In contrast, *PT4* is not induced in the DMI3 gain-of-function mutant (Takeda *et al.*, 2012). Therefore, *PT4* is supposed to represent a second wave of genes required for arbuscule formation. Ultimately, the mechanism that regulates the induction of the second wave of genes remains elusive. However, Drissner *et al.* (2007) showed that the lysolipid lyso-phosphatidylcholine (LPC), present in root extracts of mycorrhizal plants, is able to induce an ortholog of *PT4* in potato in the absence of a fungal symbiont. LPC results from partial hydrolysis of phosphatidylcholines, a major part of plasma membranes. This gives reason to speculate whether the precise temporal induction of *PT4*, which is key for its correct localization to the PAM, could be mediated by hydrolysis of plasma membrane components.

1.3 Proteases and their inhibitors

Proteases catalyze the irreversible cleavage of their target proteins and are widely distributed among nearly all organisms (Habib & Fazili, 2007). In higher organisms, about 2% of the coding genes encode for these proteolytic enzymes (Barrett *et al.*, 2001). In particular, the genomes of the model plants *Arabidopsis* and rice contain over 800 and 600 protease-encoding genes, respectively (van der Hoorn, 2008; Garcia-Lorenzo *et al.*, 2006). To carry out their function, most proteases use a special arrangement of three amino acid residues inside their active center. The so-called catalytic triad mediates the hydrolysis of a specific peptide bond within their protein substrate (Hedstrom, 2002). Based on their catalytic residues, proteases are grouped into four major classes, as there are cysteine-, serine-, aspartic-,

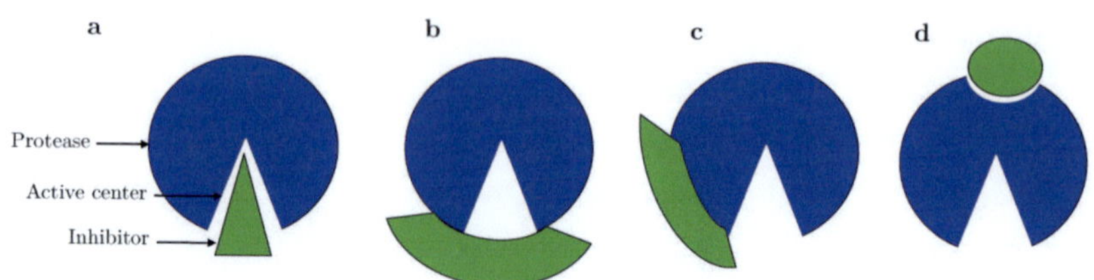

Figure 1.6: Schematic diagram of the different modes of protease inhibition. **(a)** Canonical standard mechanism according to Laskowski (1968) by direct blockage of the active center. **(b)** Indirect blockage of the active center. **(c)** Adjacent and or exosite binding. **(d)** Allosteric inhibition. Modified from Habib & Fazili (2007).

and metalloproteases (Dunn, 2001). Within those catalytic classes, they are further grouped based on sequence homology and their molecular structure. According to those criteria, proteases are classified within the MEROPS database (Rawlings *et al.*, 2012).

Uncontrolled proteolysis can be very damaging for living systems and therefore proteolytic activity needs to be strictly regulated. Next to the transcriptional control, many proteases are stored at their subcellular destinations as inactive pro-forms. Thus, the protease can rapidly be activated by cutting off the inhibitory pro-domain (Baker *et al.*, 1993). However, the most important mechanism to control protease activity is the interaction with proteins that inhibit them (Rawlings *et al.*, 2004; Habib & Fazili, 2007). An overview of the different modes of protease inhibitors to form inactive complexes with their target proteases is shown in Figure 1.6. The most common mode is the canonical standard mechanism (Laskowski, 1968; Laskowski & Kato, 1980). Thereby, a distinct reactive site loop of the protease inhibitor enters the active center of the protease in a substrate-like manner. Hence, to fully understand the biological role of a protease, it is indispensable to have information about their natural inhibitors as well. For this reason, MEROPS included a platform for protease inhibitors to comply with this need for an integrated information source. Together with their inhibitors, proteases are key regulators of a versatility of physiological processes. Those include post-translational modifications (Berezniuk *et al.*, 2012) or the initiation and transmission of signaling cascades, for example, in apoptosis or blood clotting (Turk, 2006; Martínez *et al.*, 2012). More plant-specific proteolytic functions concern fruit ripening (van der Hoorn, 2008), hormone signaling (Li *et al.*, 2001), the mobilization of storage proteins of seeds and tubers (Grudkowska & Zagdanska, 2004) and responses to abiotic and biotic stress factors (Krüger *et al.*, 2002; Tian *et al.*, 2004; van der Hoorn, 2008; van der Linde *et al.*, 2012a).

1.3.1 Protease–protein inhibitor interactions in plant–pathogen associations

In the field of biotrophic plant associations, the interaction between proteases and inhibitors at the plant–pathogen interface have received great attention as they are instrumental in deciding the fate about compatibility or disease resistance (Ryan, 1990; Xia, 2004; Kamoun, 2006; van der Hoorn, 2008; Shindo & van der Hoorn, 2008). The following two examples shall illustrate the versatility of how protease–protein inhibitor interactions can influence the outcome of biotrophic associations.

In the *U. maydis*–maize pathosystem, the maize protease inhibitor CC9 is induced upon infection and inhibits at least five different endogenous defense-related apoplastic cysteine proteases (van der Linde *et al.*, 2012a; van der Linde *et al.*, 2012b). This inhibition is necessary for pathogenicity as silencing of CC9 led to a strongly reduced fungal colonization. Furthermore, the expression of CC9 seems to be actively promoted by the fungus, as CC9 is not induced by infection with a *U. maydis pep1* deletion strain, that results in an incompatible interaction and elicits a defense response (Hemetsberger *et al.*, 2012). It is therefore assumed that the protease inhibitor CC9 is an endogenous compatibility factor, essential to suppress host immunity and to cause disease (van der Linde *et al.*, 2012a). In addition, recent studies identified another player in this scene as the secreted *U. maydis* protease inhibitor Pit2 also targets several of the CC9 inhibited cysteine proteases (Mueller *et al.*, 2013). The reason for this dual targeting of apoplastic proteases is not clear yet, however, the authors argue for spatial and temporal reasons given by the differential induction of the inhibitors during the biotrophic interaction (Mueller *et al.*, 2013).

Another interesting example of protease–protein inhibitor interactions evolves around the secreted cysteine protease Rcr3 from tomato (Krüger *et al.*, 2002). The secreted protein effector Avr2 from the fungus *Cladosporium fulvum*, but also the effector GrVAP1 from the nematode *Globodera rostochiensis* bind and inhibit Rcr3 (Rooney *et al.*, 2005; Lozano-Torres *et al.*, 2012). However, in contrast to other effectors from *Phytophtora infestans* that also target Rcr3 (Song *et al.*, 2009), the Avr2-Rcr3 and GrVAP1-Rcr3 complexes elicit a hypersensitive response, but only in tomato plants expressing the extracellular immune receptor Cf-2. This indicates that Cf-2 guards the virulence target Rcr3 and provides disease resistance upon recognition of the modifications produced on Rcr3 by interaction with Avr2 and GrVAP1 (Krüger *et al.*, 2002; Rooney *et al.*, 2005; Lozano-Torres *et al.*, 2012), whereas the modifications on Rcr3 when interacting with the *P. infestans* effectors are not monitored by Cf-2 (Song *et al.*, 2009).

1.3.2 Kunitz protease inhibitors

The proteins KPI106 and KPI104, which are the main subjects of this work, belong to the class of Kunitz protease inhibitors (KPI), as they contain a C-terminal soybean trypsin inhibitor domain (PF00197). This class of inhibitors is termed according to M. Kunitz, who identified the first member of this family in soybean meal and showed its inhibitory activity against trypsin (Kunitz, 1946; Kunitz, 1947a; Kunitz, 1947a). The protein was thus referred to as soybean trypsin inhibitor (STI; EMBL database: AB112033.1) and its interaction with trypsin became a model system to study the mechanisms of protease inhibition and protease–protein inhibitor interactions (Blow & Sweet, 1974; Sweet *et al.*, 1974; Song & Suh, 1998).

By now, based on the characteristic of single Kunitz domains to fold into stable proteins, KPIs became of great interest in the medical field with regard to the production of engineered pharmaceutical drugs (Nixon & Wood, 2006). A deregulated proteolytic function is often attached with pathological conditions, such as cancer and cardiovascular disorders (Drag & Salvesen, 2010; Scott & Taggart, 2010) and the control of aberrant proteolysis by engineered protease inhibitors provides an opportunity for therapeutic intervention (Nixon & Wood, 2006). Furthermore, KPIs often occur in venoms of some invertebrates, for example in cnidarians (Schweitz *et al.*, 1995; Minagawa *et al.*, 2008), in the mollusc cone snails (Bayrhuber *et al.*, 2005; Dy *et al.*, 2006), in arachnids (Yuan *et al.*, 2008; Zhao *et al.*, 2011), or in snakes (Strydom, 1973; Župunski *et al.*, 2003). In those toxins, the KPIs either act as protease inhibitors or have an ion channel blocking activity, which, by the virtue of the effects on predator animals, are essential for survival of those toxin-producing organisms (Fry *et al.*, 2009).

In the following, focus will be narrowed to plant Kunitz protease inhibitors. Most of the so far characterized plant KPIs were isolated from the protein-rich seeds of legumes (Birk, 2003). Based on their inhibiting preference towards serine proteases, a molecular mass between 18 and 22 kDa, and their ternary structure being predominantly stabilized by two disulfide bonds, those leguminous KPIs were included into the MEROPS I3 family (Rawlings *et al.*, 2004; Oliva *et al.*, 2010). The biological roles of seed-located KPIs are either to serve as storage proteins, and thus providing an important nitrogen source for germination (Wilson *et al.*, 1988; de Souza Cândido *et al.*, 2011), or in the combat of herbivore consumers by interfering with digestive proteases (Srinivasan *et al.*, 2005). In rice and some other monocots, seed-located KPIs that exhibit an additional inhibitory activity against endogenous α-amylases were identified. A possible role in the prevention of precocious germination is proposed (Mundy *et al.*, 1983; Nielsen *et al.*, 2004; Yamasaki *et al.*, 2006). Furthermore, KPIs are induced in leaves of various plant species upon wounding or by herbivore challenge (for example: Green & Ryan, 1972; Haruta *et al.*, 2001; Jiménez *et al.*, 2008; Major & Constabel, 2008; Oliva *et al.*, 2011), as well as during plant–pathogen interactions (for example: Valueva *et al.*, 1998; Li *et al.*, 2008; Huang *et al.*, 2010; Odeny *et al.*, 2010). Among them, KTI1 of *A. thaliana,* that is induced in response to culture filtrates from *Erwinia carotovora* (Li *et al.*, 2008). Moreover, KPI-encoding genes were up-regulated in some legumes when engaged in root nodule symbiosis (Manen *et al.*, 1991; Lievens *et al.*, 2004), suggesting a specific function in mutualistic plant-microbe associations.

1.3.3 Mycorrhizal-induced Kunitz inhibitors in *M. truncatula*

The contigs TC106351 and TC100804 of the *M. truncatula* DFCI Gene Index (MtGI) encode for two putative Kunitz protease inhibitors named *KPI106* and *KPI104*, respectively. Recent studies performed in our group revealed the induction of *KPI106* and *KPI104* in response to diffusible molecules of the AM fungus *R. irregularis* as well as during early stages of mycorrhizal symbiosis (Kuhn *et al.*, 2010). Kuhn *et al.* (2010) further showed that *KPI106* is induced in *dmi2* mutants, which indicates that the induction of *KPI106* is not dependent on the common signaling pathway. In addition, previous transcriptomic studies revealed that expression of

KPI106 and *KPI104* is also maintained during later stages (Liu *et al.*, 2003; Wulf *et al.*, 2003; Grunwald *et al.*, 2004; Hohnjec *et al.*, 2005; Liu *et al.*, 2007; Guether *et al.*, 2009a; Gomez *et al.*, 2009; Gaude *et al.*, 2012), suggesting their importance throughout the entire symbiosis. In this work, the question about the distinct biological roles of KPI106 and KPI104 in arbuscular mycorrhizal symbiosis shall be addressed.

1.3.4 Proteases in arbuscular mycorrhizal symbiosis

Independent transcriptome analyses revealed that next to the up-regulation of protease inhibitor genes, mainly three types of proteases are specifically induced during mycorrhizal symbiosis. Those include papain-like cysteine proteases, subtilases and serine carboxypeptidases (Liu *et al.*, 2003; Guimil *et al.*, 2005; Hohnjec *et al.*, 2005; Kistner *et al.*, 2005; Gomez *et al.*, 2009; Breuillin *et al.*, 2010; Gaude *et al.*, 2012). The subtilase family of *L. japonicus* as well as the *M. truncatula* serine carboxypeptidase 1 (SCP1) were part of recent research and shall be shortly introduced in the following sections.

1.3.4.1 The subtilase family of *L. japonicus*

Subtilases are one of the major classes of serine proteases and are characterized by the primary sequence order of their catalytic triad Asp, His, and Ser (Dodson & Wlodawer, 1998). The two *L. japonicus* subtilases SbtM1 and SbtM3 are exclusively expressed during mycorrhizal symbiosis and are transcriptionally induced by the common signaling pathway (Takeda *et al.*, 2011). *SbtM1* is expressed in cells containing intracellular hyphae as well as in cells that belong to the cell file that lies ahead of the fungal colonization front (Takeda *et al.*, 2009; Takeda *et al.*, 2012). Subcellular localization of a fusion protein composed of the SbtM1 signal peptide and the Venus fluorescent protein, by expression under the native *SbtM1* promoter, indicated that SbtM1 is secreted into the apoplast as well as into the PAS (Takeda *et al.*, 2009). Furthermore, silencing of either of SbtM1 or SbtM3 caused a decrease in intraradical hyphal and arbuscule abundances. It is thus assumed that both subtilases would cleave a substrate located at the plant–fungal interface and that this cleavage is crucial for the formation of arbuscules (Takeda *et al.*, 2009).

1.3.4.2 *M. truncatula* serine carboxypeptidase 1

Serine carboxypeptidases represent another major group of serine proteases and contain a catalytic triad in the sequence order Ser, Asp, His (van der Hoorn, 2008). An interesting feature about SCPs is the variety of the reactions they catalyze. In the *Arabidopsis* mutant *sng1,* Lehfeldt *et al.* (2000) identified the affected gene to display all sequence features that are characteristic for SCPs. However, instead of hydrolytic activity, they found SNG1 to catalyze a trans-esterification, and thus acting as an acyltransferase. Analyzing the structure–function relationship between SCPs and SCP-like acyltransferases in *Arabidopsis,* Stehle *et al.* (2009) found that SCPs exhibiting hydrolytic activity contain the characteristic pentapeptide –GESYA– around their catalytic Ser residue. In contrast, SCP-like acyltransferases reveal –GDSYS– instead, suggesting that they evolved by neofunctionalization from a hydrolytic ancestor (Stehle *et al.*, 2009).

In *M. truncatula,* the contig TC106954 (MtGI) encodes a predicted serine carboxypeptidase, named *SCP1*, which has been identified in many transcriptome analyses to be induced during mycorrhizal symbiosis (Liu *et al.*, 2003; Hohnjec *et al.*, 2005; Liu *et al.*, 2007; Floss *et al.*, 2008b; Gomez *et al.*, 2009; Gaude *et al.*, 2012). *SCP1* is mainly expressed in arbusculated and adjacent cortical cells (Liu *et al.*, 2003; Gomez *et al.*, 2009; Pumplin & Harrison, 2009). Furthermore, similarly to *SbtM1*, *SCP1* is also expressed in regions of outer cortical cells in which the underlying inner cortex is colonized (Liu *et al.*, 2003). Pumplin & Harrison (2009) used a *SCP1*promoter:GFP reporter to co-localize with different organelle markers to monitor cellular dynamics in cortical cells during arbuscule development. A closer inspection of these stunning images shows that the GFP fluorescence is stronger in the developing and mature arbuscules in contrast to collapsing arbuscules. This suggests that SCP1 is rather required during arbuscule development than for arbuscule turnover. The specific function of SCP1 has not been analyzed so far. However, the results of this work reveal that SCP1 is crucial for arbuscule development as silencing of SCP1 and homologous serine carboxypeptidases led to a defective arbuscule phenotype.

1.4 Aim of this study

The crucial role of protease–protein inhibitor interactions in plant–pathogen associations is evident, and the number of examples illustrating this is steadily increasing. However, knowledge about the functional role of proteases and their inhibitors in symbiotic associations is limited. *KPI106* and *KPI104* belong to a small family of *M. truncatula* genes that encode putative Kunitz protease inhibitors and are exclusively expressed during mycorrhizal symbiosis. The question about their biological role in arbuscular mycorrhizal symbiosis will be addressed in this work. By means of *in silico* analyses and structure modeling, the putative Kunitz inhibitors will be examined for typical sequence features in order to integrate them into the population of the characterized plant KPIs and to further assign their predicted biochemical function. As a previously employed loss-of-function approach to silence *KPI106* and *KPI104* did not result in an effect on the symbiosis, their function shall be further investigated by deregulation of their spatio-temporal expression. Using directed and non-directed approaches, putative target proteases of the mycorrhizal-induced Kunitz inhibitors shall be unraveled. Thereby, this study elucidates that several members of a family of *M. truncatula* serine carboxypeptidases interact with individual members of the Kunitz family. The arising question whether putative tandem interactions between the KPIs and the SCP proteins might occur will be discussed in this work. Finally, by means of a reverse genetics approach the biological role of the KPI-interacting SCPs in arbuscular mycorrhizal symbiosis shall be addressed.

CHAPTER 2

RESULTS

2.1 A family of Kunitz protease inhibitors is induced in *M. truncatula* during AM symbiosis

A detailed sequence analysis of *KPI106* and *KPI104* within the *M. truncatula* genome (hapmap 3.5) revealed the presence of a large family of 25 genes encoding for putative Kunitz protease inhibitors. Figure 2.1 illustrates the resulting phylogenetic tree based on cDNA sequences, in which the STI of *Glycine max* (Kunitz, 1946; EMBL database: *AB112033.1*) was included as outgroup. The tree shows that *KPI106* and *KPI104* cluster within a distinct subfamily that is composed of six members. *In silico* analysis of expression using the *M. truncatula* Gene Atlas (MtGEA) showed that only those genes within this cluster are mycorrhizal-induced. This is consistent with Grunwald *et al.* (2004), revealing the induction

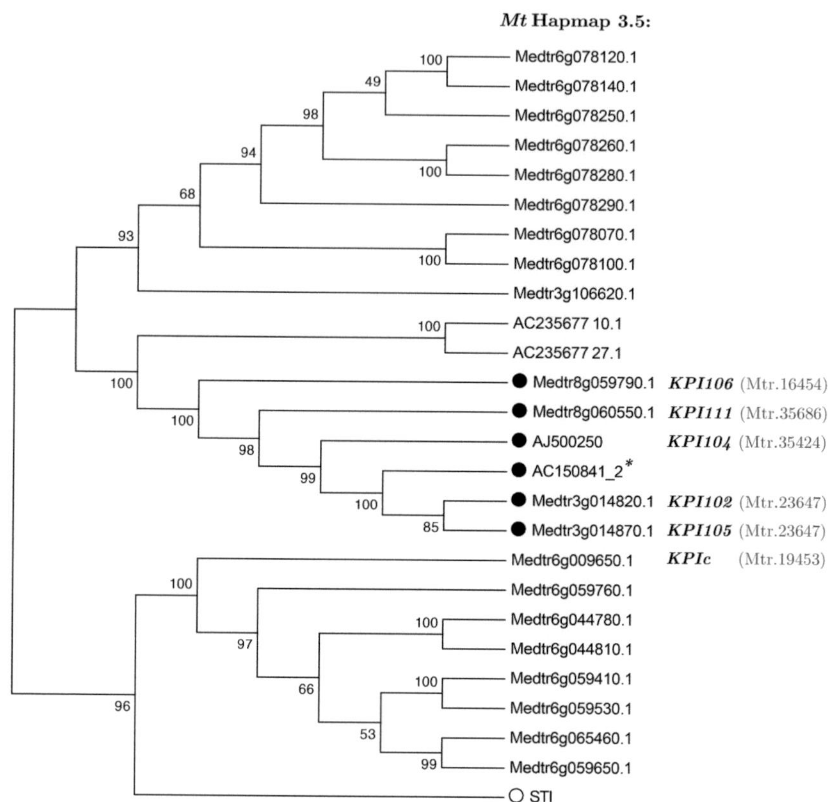

Figure 2.1: Phylogenetic tree of Kunitz inhibitor genes in the *M. truncatula* genome. Blast searches of *KPI106* and *KPI104* against the hapmap 3.5 database revealed 25 Kunitz inhibitor genes within the *M. truncatula* genome, whereas *KPI106* and *KPI104* cluster in a predicted mycorrhizal-induced subfamily (black dots; MtGEA probeset IDs in gray). *AC158047_2** is a partial C-terminal sequence and was therefore not included in further experiments. *KPIc* was predicted as not being expressed during AM symbiosis and will therefore be used as control in further experiments. The Kunitz inhibitor *STI* of *G. max* was used as outgroup. The tree is based on cDNA sequences and was constructed using the Neighbor-Joining algorithm and numbers represent bootstrap supported values.

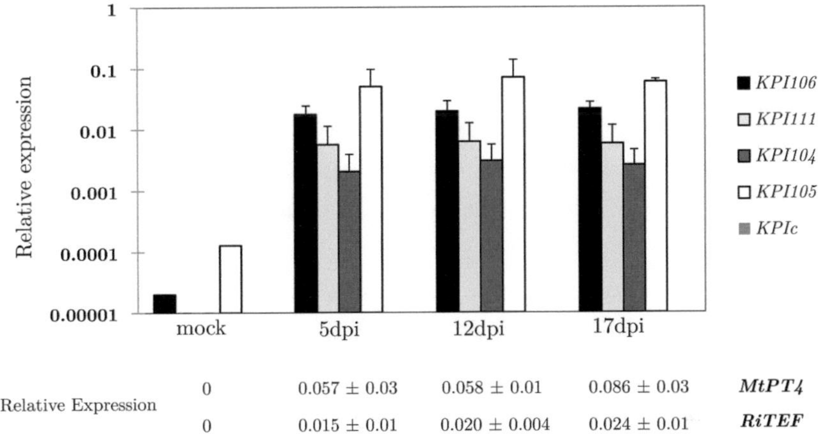

Figure 2.2: Expression of selected *KPIs* in *R. irregularis* colonized *M. truncatula* hairy roots. All predicted mycorrhizal-induced *KPIs* were expressed in *R. irregularis* colonized *M. truncatula* hairy roots, whereas the control inhibitor *KPIc* was not expressed. Fungal abundance within the root samples are indicated by quantification of the *R. irregularis* houskeeping gene *RiTEF*. The relative expression levels of the *M. truncatula* symbiotic phosphate transporter *PT4* indicate the activity of mycorrhizal symbiosis. All relative expression values are represented by means of three independent biological replicates and standard deviations are indicated with error bars and numbers. All values were normalized to *M. truncatula TEF*.

of five putative Kunitz protease inhibitor genes in *M. truncatula* roots when colonized by *G. mossae*. Following the nomenclature adopted for *KPI106* and *KPI104*, the other mycorrhizal-induced KPIs were named according to their contig annotation of the MtGI as *KP111* (TC111102), *KPI102* (TC197802) and *KPI105* (TC100805). The partial sequence of *AC150841_2* (indicated with * in Figure 2.1) could not be supported by any TC assembly, and was thus excluded from the following analyses. The putative Kunitz inhibitor named as *KPIc* is predicted (MtGEA) to be mainly expressed in *M. truncatula* seeds and not during mycorrhizal symbiosis and was thus used as a control inhibitor in further experiments.

Next, the expression pattern of those mycorrhizal-induced *KPIs* as well as *KPIc* were analyzed over a time course of 5 dpi to 17 dpi in the *M. truncatula/R. irregularis* system. Results showed that the predicted mycorrhizal-induced *KPI* genes were fully induced at 5 dpi and expression maintains up to 17 dpi, with *KPI105* having higher transcript levels throughout all time points analyzed (Figure 2.2). In fact, *KPI105* and *KPI106* were expressed at low levels in non-mycorrhizal roots whereas *KPI104* and *KPI111* were not. Furthermore, in accordance to the MtGEA, *KPIc* was not expressed in colonized or non-colonized *M. truncatula* roots. The mycorrhizal status of the tested root samples was assessed by quantification of the *R. irregularis* housekeeping gene *RiTEF*, corresponding to the amount of fungus, as well as the *M. truncatula* symbiotic phosphate transporter *PT4* (Harrison *et al.*, 2002; Javot *et al.*, 2007) indicative of the symbiotic activity. The results revealed that both transcripts accumulate parallel to the proceeding symbiosis.

A common feature for the classification of Kunitz inhibitors is their content in cysteines attended by the number of disulfide bonds (Rawlings *et al.*, 2004). An alignment of the amino acid sequences of the mycorrhizal-induced KPIs revealed that all proteins contain six conserved cysteines, indicative of three putative disulfide bonds (Figure 2.3a). This is in contrast to most described leguminous Kunitz inhibitors that only contain two disulfide bonds, as for example, the prominent STI from *G. max* (Kunitz, 1947b; Rawlings *et al.*, 2004; Oliva *et al.*, 2010). According to MEROPS, plant KPIs including STI (MEROPS ID: I03.001) belong to the I3A family of protease inhibitors. However, blast searches of KPI106 against MEROPS revealed that apparently all mycorrhizal-induced KPIs are classified into group I3B, that includes so far only the two protease inhibitors from *Sagittaria sagittifolia* API-A (MEROPS ID: I03.006) and API-B (MEROPS ID: I03.007). From the crystal structure of API-A, it is evident that the protein is stabilized by three disulfide bonds spanning between Cys^{34}–Cys^{89}, Cys^{139}–Cys^{148} and Cys^{141}–Cys^{144} (Bao *et al.*, 2009).

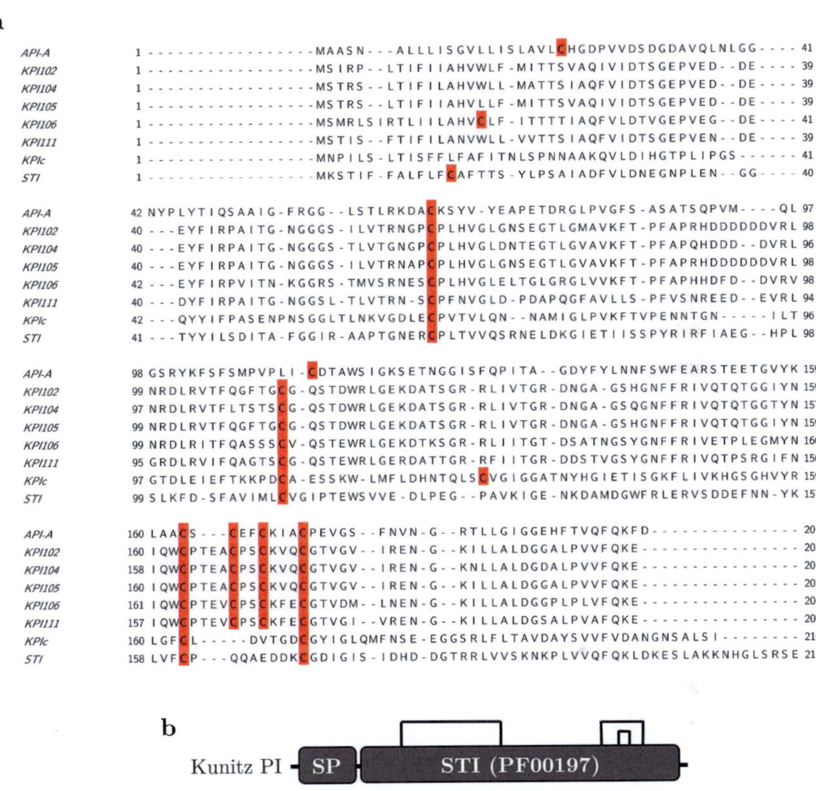

Figure 2.3: Domain structure of mycorrhizal-induced KPIs. **(a)** Amino acid alignment of mycorrhizal-induced KPIs shows conservation of six cysteine residues (red) as in the protease inhibitor API-A from *S. sagittifolia*, whose crystal structure is already solved. In contrast, STI from *G. max* and KPIc only contain four conserved cysteines. **(b)** Mycorrhizal-induced KPIs contain an N-terminal secretion signal peptide and a C-terminal Pfam PF00197 STI domain. Three putative disulfide bonds in analogy to API-A are indicated in black brackets.

Comparison of the API-A and KPI sequences showed that indeed all cysteine residues are conserved between them (Figure 2.3a), suggesting that the mycorrhizal-induced KPIs possess the same disulfide bond pattern as API-A (Figure 2.3b). Additionally, all deduced KPI protein sequences are predicted to contain an N-terminal secretion signal peptide (Signal4P; Target4P) and a soybean trypsin inhibitor domain of protease inhibitors (Pfam PF00197; Figure 2.3b).

2.2 KPI homologs in other plant species

So far, the MEROPS database only contains two members within the I3B group of plant Kunitz protease inhibitors. To investigate the occurrence of potential homologs of the mycorrhizal-induced KPIs among other plant species, a blastx screen with the *KPI106* cDNA sequence against the NCBI database was carried out. The phylogenetic tree in Figure 2.4 highlights the best 50 hits of the identified proteins, whereas sequences within the genera of *Medicago* were not considered. In analogy to the tree of Figure 2.1, the STI was used as outgroup. As expected, the closest homologs of KPI106 were found among other legume species including chickpea (*Cicer*), soybean (*Glycine*) and *Lotus* as another representative of clover. In addition, several larger families of KPI homologs were identified in the more distant related genera of grapevine (*Vitis*) and poplar (*Populus*).

With attention to the distribution of cysteines among the identified homologs, the amino acid sequences of their respective C-termini were aligned (Figure 2.5). Results showed that, except for two Kunitz inhibitors of the genus of *Cucumis*, all other identified homologs contain the conserved four cysteines in this area, analogous to KPI106, but in contrast to STI that only has two conserved C-terminal cysteines. This would indicate that the triple disulfide bond feature is more common in plants than expected by

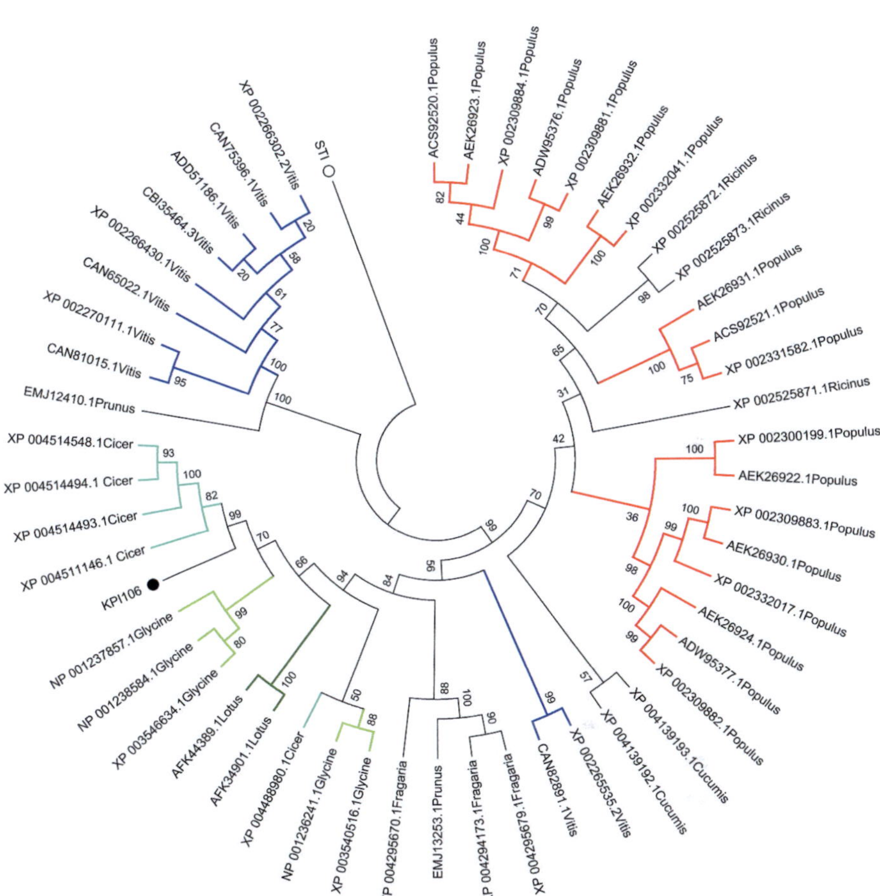

Figure 2.4: Homologs of mycorrhizal-induced KPIs. The phylogenetic tree contains the best 50 hits of a blastx screen of the *KPI106* cDNA sequence against the NCBI database. Most of the KPI106 (black dot) homologs are found in other legume species including chickpea (*Cicer*), soybean (*Glycine*) and *Lotus* (all in greenish colors). Two larger groups of homologs are present in poplar (red) and grapevine (blue). The tree is based on protein sequences and was constructed using the Neighbor-Joining algorithm. The conventional STI of *G. max* (empty black dot) was used as outgroup, and numbers represent bootstrap-supported values.

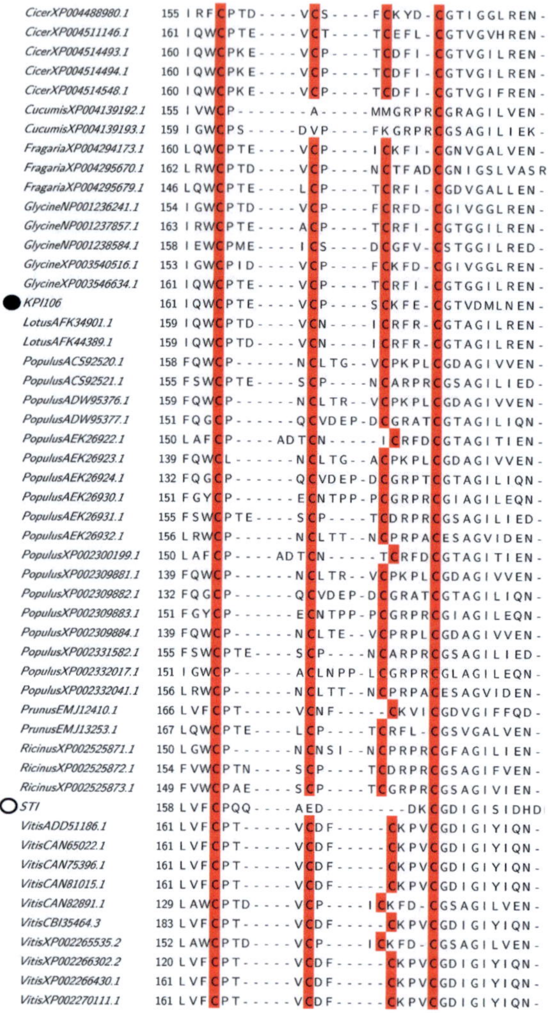

Figure 2.5: Disulfide bond pattern of homologous KPI proteins. Amino acid alignment of the C-terminal parts of the best 50 identified KPI106 homologs. The overall majority of the homologous KPI proteins contains four conserved cysteine residues in their respective C-termini, as does KPI106 (black dot). In contrast, STI (black circle, bottom) only contains two conserved cysteine residues in this area.

the I3B entry in the MEROPS database, assumed that those six conserved cysteines lead to the formation of three disulfide bonds in all cases. It further suggests that this feature might be related to a distinct function when compared to KPIs of leguminous seeds, of which the overall majority does not hold the triple disulfide bond pattern (Oliva *et al.*, 2010).

2.3 Overexpression of the KPIs leads to a mycorrhizal phenotype

To functionally characterize the mycorrhizal-induced KPIs, an RNAi approach to silence *KPI106* and *KPI104* was already performed within frame of my Diploma thesis. However, reduction of the single *KPI106* and *KPI104* transcripts did not result in an obvious effect on the symbiosis (see Figures S1 and S2).

In this work, to further investigate the function of these proteins, an overexpression approach to deregulate the spatio-temporal expression of *KPI106* and *KPI104* by constitutive expression under the cauliflower mosaic virus 35S promoter was carried out. The overexpression was confirmed by PCR on cDNA of non-mycorrhizal roots, revealing that *KPI106* and *KPI104* transcripts are present in all analyzed samples, in contrast to non-mycorrhizal control roots carrying an empty vector (Figure 2.6a). Microscopical examination of the WGA-fluorescein stained mycorrhizal roots revealed an altered mycorrhizal morphology with a remarkable high amount of malformed arbuscules in KPI overexpressing roots, in contrast to the control roots (Figure 2.6b–d). Quantification of fungal colonization confirmed these morphological observations, and showed that the overall abundance of mature arbuscules in the overexpression roots was significantly reduced when compared to the control roots (Figure 2.7). In addition, whereas frequency and intensity of the mycorrhizal colonization

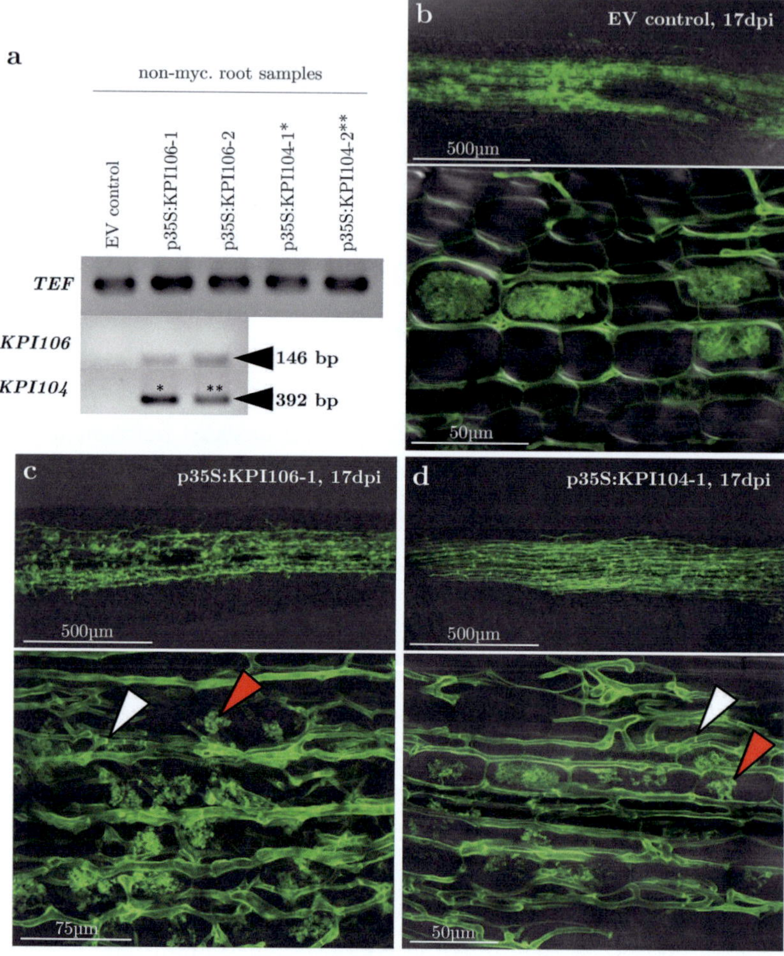

Figure 2.6: KPI106 and KPI104 overexpression leads to a mycorrhizal phenotype. **(a)** *KPI106* and *KPI104* transcripts were detectable in cDNA samples of non-mycorrhizal overexpression lines, but not in the control line carrying an empty vector. **(b–d)** Confocal microscopy of WGA-fluorescein stained mycorrhizal roots at 17 dpi. In KPI overexpressing roots, many malformed arbuscules (red arrowheads) were visible and the intraradical hyphae were often septated (white arrowheads) when compared to the control roots. Morphological observations in the mycorrhizal hairy root lines p35S:KPI106-2 and p35S:KPI104-2 did not differ from the ones presented in this figure. Images represent maximum projections of z-stacks.

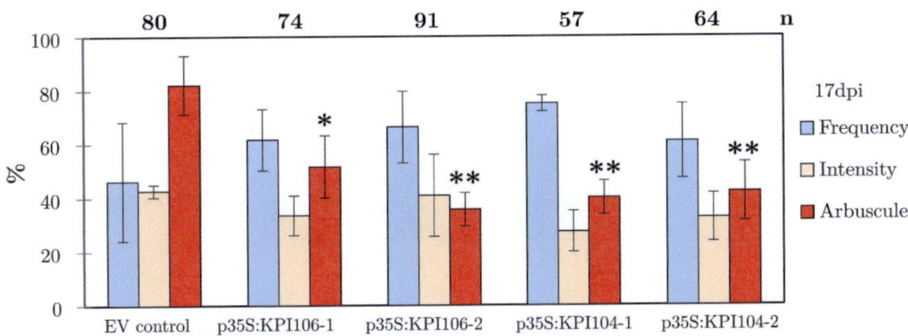

Figure 2.7: Quantification of mycorrhizal colonization in KPI overexpressing roots. Quantification of mycorrhizal colonization at 17 dpi revealed a significant reduced number of mature arbuscules (red columns) in the overexpression lines, whereas the frequency (blue columns) and intensity of colonization (orange columns) within the root fragments did not differ from those in the control roots. Bars represent means of the number of root fragments, and errors indicate standard deviations. n: number of root fragments; **: p-value < 0.01; *: p-value < 0.05.

within the root fragments was similar in all lines, the fungal hyphae in the overexpression lines were often septated, which is a morphological signature for cell death in AM fungi (Bonfante-Fasolo, 1984).

To further dissect this arbuscule phenotype, the transcript abundances of the mycorrhizal markers *PT4* (Harrison *et al.*, 2002; Javot *et al.*, 2007) and the symbiotic monosaccharide transporter *MST2* of *R. irregularis* (Helber *et al.*, 2011) were quantified at 17 dpi, consistent with the time point of mycorrhizal quantification. Results revealed that both genes are appropriately expressed in the *KPI* overexpression roots when compared to the control roots (Figure 2.8). Interestingly, the *KPI104* transcript was almost 100-fold more abundant in overexpression line 1 (p35S:KPI104-1; Figure 2.8) compared to line 2 (p35S:KPI104-2), indicating that in this root line the T-DNA integration might have occurred into a more transcriptionally active locus or that several T-DNA copies were integrated into

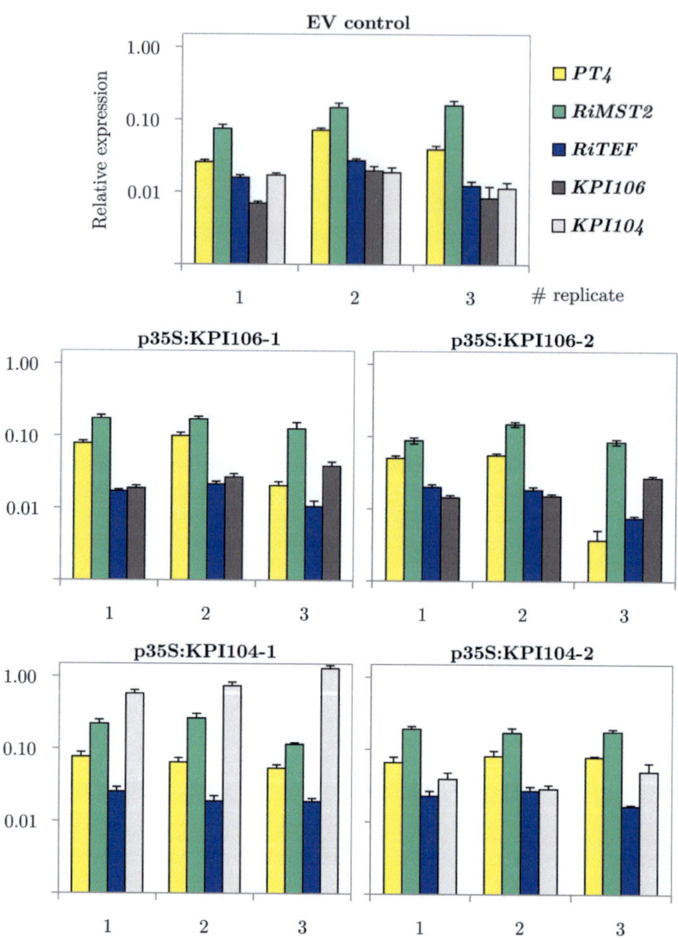

Figure 2.8: Dissecting the KPI overexpression phenotype: Quantification of mycorrhizal markers. Transcript abundances of three independent biological replicates (indicated with #) of each hairy root line at 17 dpi. Expression levels of *KPI106* and *KPI104* in the overexpression lines did not exceed their native expression within the control roots at 17 dpi. The mycorrhizal markers *PT4* and *RiMST2* are appropriately expressed in all tested samples, assigning the arbuscule phenotype to be *PT4* independent. *RiMST2* and the *R. irregularis* housekeeping gene *RiTEF* are comparable in all lines, indicative of fungal viability. Errors represent standard deviations of three technical replicates and bars show relative expression values normalized to *MtTEF*, or *RiTEF* in case of *RiMST2*.

the genome. Additionally, the mycorrhizal markers in p35S:KPI104-1 are also higher expressed. However, compared to the transcript abundance of *RiTEF*, the induction of those may rather be due to the higher fungal abundance within those roots. Moreover, Figure 2.8 shows that constitutive expression of the *KPIs* under p35s does not exceed their native expression in the mycorrhizal empty vector control roots at 17 dpi, indicating that it is not the overexpression *per se* but the disruption of their spatio-temporal expression what causes the mycorrhizal phenotype. Gutjahr & Parniske (2013) and Groth *et al.* (2013) propose that arbuscule development is associated with two distinct waves of gene induction *in planta*. Therefore arbuscule phenotypes were grouped as either *PT4* dependent or independent, according to the absence or presence of the *PT4* transcript in the respective colonized mutants. The arbuscule phenotype produced by KPI overexpression can thus be assigned as *PT4* independent (Figure 2.8). Furthermore, the arbuscule phenotype seems not to be caused by a lower fungal viability, as expression of *R. irregularis MST2* and *TEF* in the KPI overexpression roots did not significantly differ from those in the empty vector control roots, consistent with the microscopically determined colonization levels (see Figure 2.7).

2.4 KPI106 and KPI104 interact with a *M. truncatula* cysteine protease

To identify potential targets of KPI106 and KPI104, a non-directed Gal4-based yeast two-hybrid assay was carried out as part of my Diploma Thesis. Thereby, the KPI106 bait protein was used to screen a cDNA library of mycorrhizal *M. truncatula* roots and 156 positive clones were identified. Thereby, the intensity of the identified interactions, indicated by expression of the α-Gal reporter, as well as the presence of an N-terminal secretion signal peptide of the interacting proteins were used as selection criteria for putative interaction partners. Results revealed a strong interaction between KPI106 and a clone encoding 150 aa of the carboxy-terminal part of a secreted *M. truncatula* cysteine protease (CP; see also Table S1).

To follow up these results, the complete *CP* open-reading frame could be mapped to the MtGI contig TC133093. Under the contigs IMGAG 50810_1.1 and 50810_1.2, the *CP* sequence can be accessed in the genome assembly of the J. Craig Venter Institute (Figure 2.9a). According to the MtGEA (Mtr. 10398.1), *CP* is predicted to be expressed ubiquitously in various plant tissues. To integrate the expression pattern of *CP* with the proceeding mycorrhizal symbiosis, *CP* transcript levels were analyzed in the same cDNA samples used for the time course analysis of the *KPIs* (see Figure 2.2). Results showed that *CP* is highest expressed in mock inoculated roots, consistent with the prediction of the MtGEA. Furthermore, the results revealed that *CP* expression slightly differs between the analyzed mycorrhizal time points (Figure 2.9b).

Next, the full-length CP protein, without its signal peptide was tested for interaction with KPI106 as well as with KPI104, and in both cases a positive interaction was found (Figure 2.9c). To verify these interactions, a subsequent *in vitro* pulldown experiment was performed. Therefore, the proteins were heterologously expressed in the *E. coli* BL21 strain. The amino-terminus of CP was thereby fused to GST, whereas KPI106 and KPI104 were expressed as N-terminal 6×His-tagged versions. The α-His probed western blot in Figure 2.9d shows that CP-GST was able to successfully pulldown 6×His-KPI106 and 6×His-KPI104 out of an *E. coli* crude extract, whereas in the negative controls no unspecific binding of the 6×His-tagged KPIs to the GST itself or the agarose matrix was observed (see Figure S3 for the according α-GST western blots). The weak bands in the 6×His-KPI106 input and pulldown lanes result from excessive aggregation of 6×His-KPI106 in inclusion bodies, and

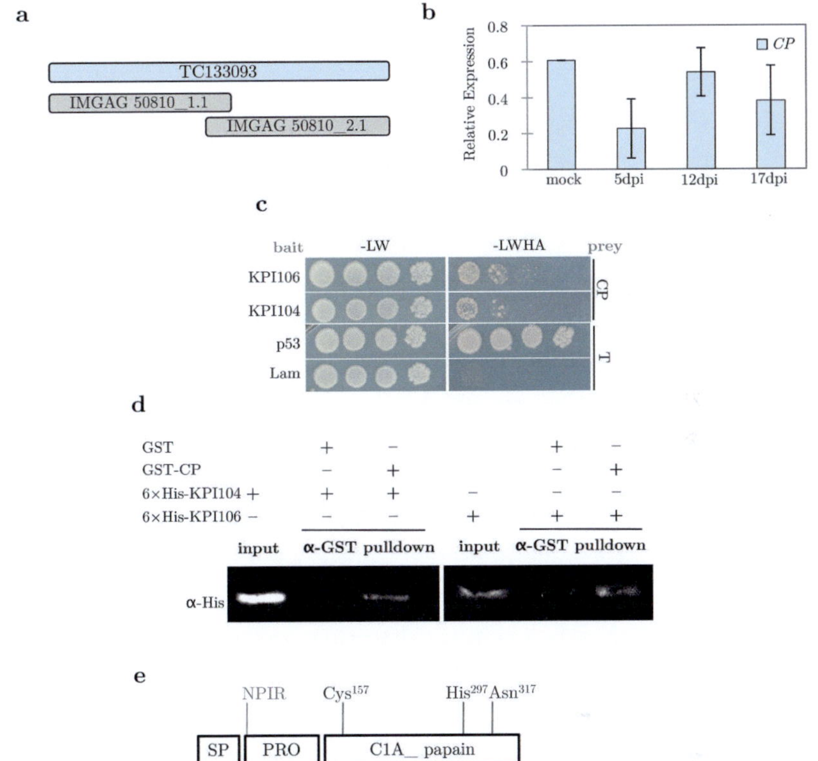

Figure 2.9: Interaction of KPI106 and KPI104 with a *M. truncatula* cysteine protease. (a) The full length cDNA sequence of *CP* was found on TC133093 in the MtGI and on IMGAG_50810_1.1/2. **(b)** *CP* is not mycorrhizal-induced and was expressed at highest in mock-treated roots. The *CP* expression pattern differed at distinct time points during mycorrhizal symbiosis. *CP* expression was normalized to *TEF* and error bars are standard deviations of three independent biological replicates. **(c)** KPI106 and KPI104 show a strong interaction with CP in the yeast two-hybrid system. **(d)** *In vitro* GST pull-down of 6×His-KPI106 and 6×His-KPI104 by GST-CP out of an *E. coli* crude extract. The α-His western blot shows bands of 6×His-KPI106 and 6×His-KPI104 in lanes were GST-CP was used as input, in contrast to the negative controls using free GST as bait. **(e)** *In silico* analysis of the CP protein sequence. The N-terminal secretion signal peptide is followed by an autoinhibitory pro-domain carrying the –NPIR– vacuolar sorting motif and a C1A domain of papain proteases including the predicted catalytic residues Cys^{157}, His^{297} and Asn^{317}.

thus less protein was available for the pulldown experiment. The CP protein sequence is predicted to contain an N-terminal secretion signal peptide, followed by an autoinhibitory pro-domain and a C1A domain of papain proteases harboring the catalytic residues Cys^{157}, His^{297} and Asn^{317} (Figure 2.9e). However, a more detailed sequence analysis revealed that the CP autoinhibitory pro-domain contains the –NPIR– vacuolar sorting motif (Ahmed *et al.*, 2000; Richau *et al.*, 2012), and therefore the interaction of CP with KPI106 and KPI104 taking place in the apoplast seems unlikely. However, it cannot be excluded that the proteins interact *in vivo* as they all pass through the secretory pathway.

In parallel, to ascertain whether CP played a possible role in AM symbiosis, an RNAi approach was carried out. Figure 2.10a shows successful silencing of *CP* in two *M. truncatula* root lines by quantitative real-time PCR. In mycorrhizal *CP*-silenced roots, many arbuscules were irregularly shaped (Figure 2.10b) and analysis of arbuscule abundances confirmed this observation as significantly less mature arbuscules were quantified in *CP*-silenced roots when compared to the empty vector control roots (Figure 2.10c). Whereas the frequency of colonization was comparable, the overall intensity of mycorrhizal colonization within the root fragments was lower in the *CP*-silenced roots when compared to the control. This was not the case in the phenotype produced by the overexpression of *KPI106* and *KPI104*.

Analysis of the mycorrhizal markers at 17 dpi showed comparable amounts of the *PT4*, *RiMST2* and *RiTEF* transcripts in *CP*-silenced and control roots (Figure 2.11a), that was also observed in colonized *KPI106* and *KPI104* overexpressing roots at 17 dpi (compare Figure 2.8). In contrast, *CP*-silenced roots showed morphological abnormalities including a brownish color and a bushy growth when compared to the control roots (Figure 2.11b–c). This indicates that the mycorrhizal phenotype produced by *CP*-silencing might be non-specific.

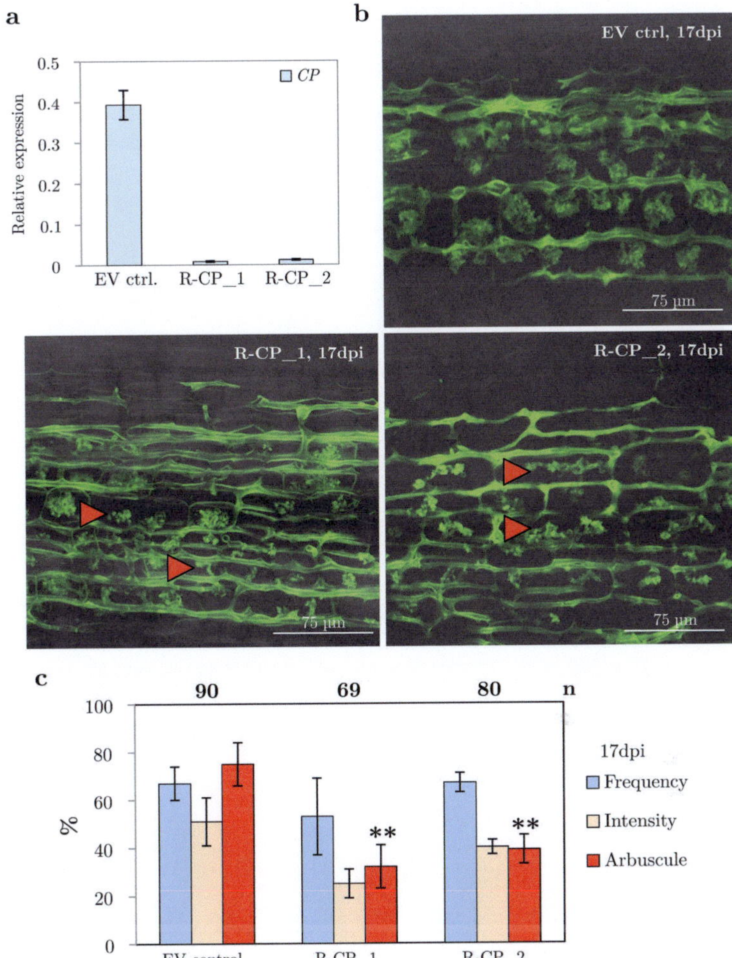

Figure 2.10: Silencing of *CP* in *M. truncatula* hairy roots. **(a)** qPCR analysis confirmed silencing of *CP* (normalized to *TEF*) in two *M. truncatula* hairy root lines. Error bars represent means of three technical replicates. **(b)** Arbuscules at 17 dpi in RCP roots were irregularly shaped (red arrow heads) when compared to the control roots (green: WGA-fluorescein). Images are maximum projections of z-stacks. **(c)** Quantification of mycorrhizal colonization revealed a significant less number of mature arbuscules (red column) and a lower colonization intensity of the root fragments (orange column) in both RCP lines, in contrast to the control roots. **: p-value < 0.01.

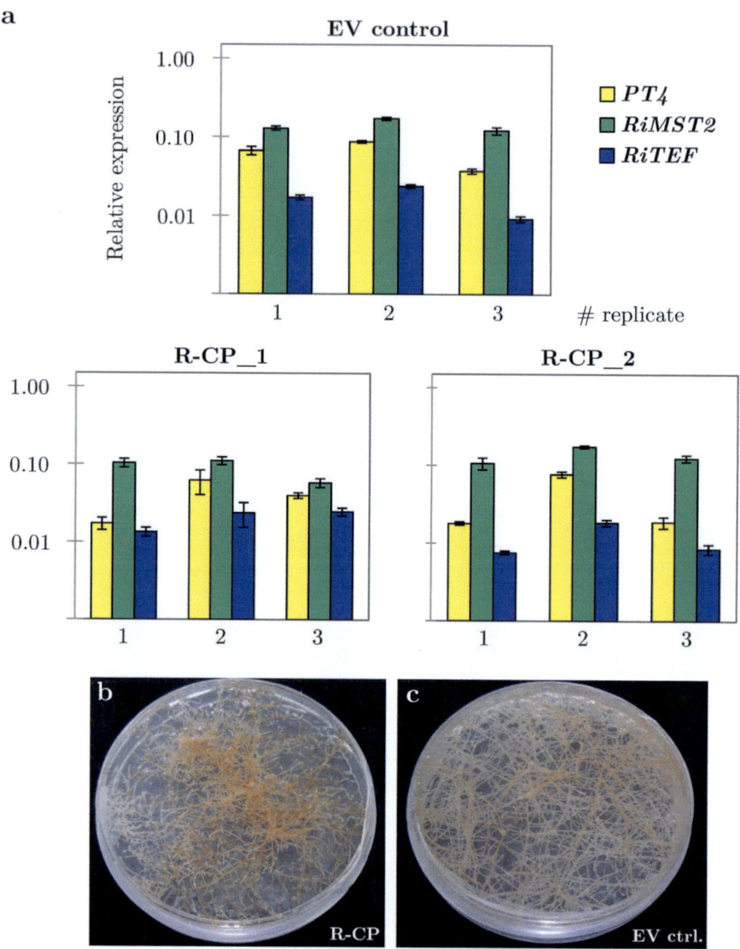

Figure 2.11: The phenotype produced by *CP*-silencing is non-specific. (a) Mycorrhizal markers *PT4* and *RiMST2* as well as the fungal housekeeping gene *RiTEF* showed comparable expression levels in the *CP*-silenced and empty vector control lines at 17 dpi. Bars represent relative expression levels in three independent biological replicates (indicated with #) of each line. Errors represent standard deviations of three technical replicates. Values were normalized to *MtTEF*, or *RiTEF* in case of *RiMST2*. **(b)** *CP*-silenced roots on MS phytagel plates for propagation showed a brownish coloring and a bushy growth compared to control roots, suggesting that the mycorrhizal phenotype of *CP*-silencing might be non-specific.

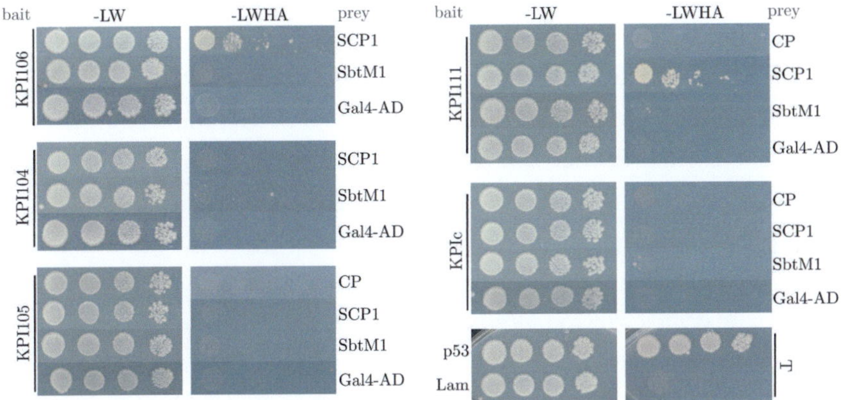

Figure 2.12: Individual interaction affinities between KPIs and mycorrhizal-induced proteases. The Kunitz inhibitors showed individual interaction patterns with the mycorrhizal-induced proteases SCP1 and SbtM1. Controls include the interactions of the KPIbait proteins with the Gal4 activation domain (Gal4-AD) and the previously identified CP (see Figure 2.9a for CP interaction with KPI106 and KPI104). KPIc as control Kunitz inhibitor did not interact with any of the tested proteases, confirming the specificity of the identified interactions. Interaction of the yeast proteins p53–T and Lam–T served as positive and negative control, respectively.

2.5 KPIs show individual interaction patterns with mycorrhizal-induced proteases

Coupled with the induction of protease inhibitors, some proteases were reported to be exclusively expressed during mycorrhizal symbiosis (summarized in Takeda *et al.*, 2007). Among them, the currently investigated *M. truncatula* serine carboxypeptidase SCP1 (Liu *et al.*, 2003; Pumplin & Harrison, 2009) and the *L. japonicus* subtilase SbtM1 (Takeda *et al.*, 2009; Takeda *et al.*, 2011), which was already shown by Takeda *et al.* (2009) to be crucial for arbuscule development, were considered to display potential targets of the KPIs. To test this hypothesis, a direct yeast two-hybrid assay was performed to analyze the interactions of the KPIs with SCP1 and the *M. truncatula* closest homologue of SbtM1 (AW584611; MtGI). In accordance to the CPprey construct (see Figure 2.9c), the coding regions of SCP1 and SbtM1 without the secretion signal peptide but including the N-terminal prodomain were amplified and cloned into the prey vector. In addition, CPprey was used as internal positive control for the assay; however, in the following only the interactions of CP with KPI105 and KPI111 shall be presented, as the interactions with KPI106 and KPI104 did not differ from the results presented in Figure 2.9c. Furthermore, to confirm specificity of the interactions, all proteases were tested for interaction with the control inhibitor KPIc (Figure 2.1). KPIc is not expressed in mycorrhizal *M. truncatula* roots (compare Figure 2.2); therefore KPIc should not interact with any of the mycorrhizal-induced proteases. Moreover, all tested KPI bait proteins (expressed as Gal4 DNA-binding fusions) were tested for interaction with the Gal4 activation domain (Gal4-AD) to exclude self-activation of the reporter genes (Figure 2.12).

Results showed that, despite the high sequence similarity among the KPIs, individual interaction affinities could be observed for each one. Thus, whereas KPI106 showed a strong affinity for SCP1 and CP, KPI111 only interacted with SCP1. Moreover, KPI104 interacted with CP, whereas KPI105 did not interact with any of the tested proteases. None of the Kunitz inhibitors interacted with SbtM1. Most importantly, KPIc did not interact with any of the tested proteases confirming the specificity of the identified interactions with the mycorrhizal-induced KPIs.

In order to further characterize the interactions of the protease SCP1, an *in silico* approach was employed. With respect to the mycorrhizal phenotype observed for the KPI106 overexpression, the tandem of KPI106 and SCP1 was chosen for further analysis and results shall be presented in the following section.

2.6 KPI106 Lys^{173} is involved in the interaction with the protease SCP1

The most common operating mode of proteinaceous protease inhibitors is the canonical standard mechanism (Laskowski, 1968; Laskowski & Kato, 1980). Thereby, a single reactive-site peptide bond is used to mimic the substrate of the protease, the so-called scissile bond (Ozawa & Laskowski, 1966; Laskowski & Qasim, 2000). The corresponding residues next to the scissile bond are labeled P_1, P_2, and so on in N-terminal- as well as $P_{1'}$, $P_{2'}$, and so on in C-terminal direction. To analyze the KPI106–SCP1 tandem in more detail, three-dimensional structure models of the two proteins were constructed and *in silico* docking was performed in order to map the interaction site between the two proteins. It should be noticed that the following protein models, the *in silico* docking as well as the three-dimensional structure alignments were carried out in frame of the Bachelor thesis of Sven Heidt (2012, KIT).

The models for SCP1 and KPI106 were constructed based on homology modeling by using the structure prediction server $PHYRE^2$ that revealed the crystal structure of the protease inhibitor API-A (Bao *et al.*, 2009) as best template model for KPI106. In Figure 2.13a–b the two models show a comparable arrangement of the three characteristic disulfide bonds. Furthermore, Lys^{169} has been experimentally assigned as the P_1 residue in API-A (Bao *et al.*, 2009) that is located in an exposed position of a distinct loop formed by two of the disulfide bonds (Figure 2.13a). Accordingly, the structure model of KPI106 also contained a Lys (Lys^{173}) in an analogous position (Figure 2.13b). Furthermore, *in silico* docking with SCP1 revealed that a loop ranging from Cys^{164} to Cys^{176} of KPI106 perfectly matches the active pocket of SCP1, which harbors the predicted catalytic triad Ser^{226}, Asp^{414} and His^{466} (Figure 2.13c). Moreover, KPI106 $Lys^{173}$is in close

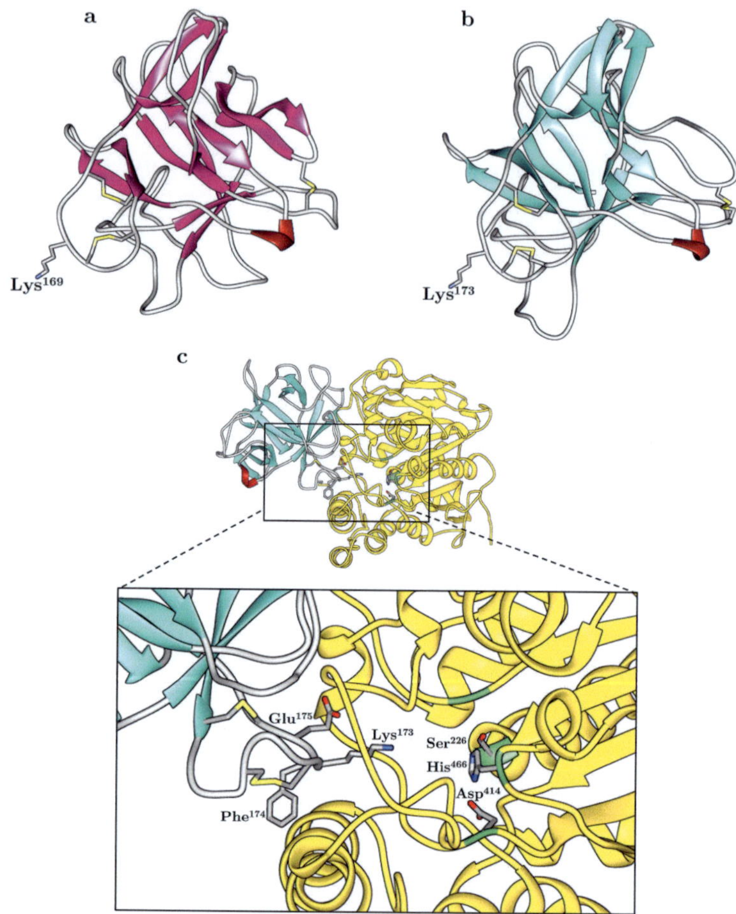

Figure 2.13: *In silico* docking of KPI106 with SCP1. The structure model of the protease inhibitor API-A **(a)** served as template to model KPI106 **(b)**. The arrangement of the three disulfide bonds (yellow) is comparable in both models and KPI106 contains a Lys (Lys^{173}) at the same position as the P_1 residue Lys^{169} of API-A. **(c)** *In silico* docking of KPI106 (turquoise) and SCP1 (yellow) shows that a distinct loop of KPI106 perfectly matches the active center of SCP1 that contains the catalytic residues Ser^{226}, Asp^{414}, His^{466}. This conformation brings Lys^{173} in close contact to the catalytic Ser^{226}of SCP1, suggesting Lys^{173} as putative P_1 of KPI106. Also highlighted are the putative $P_{1'}$ Phe^{174} and $P_{2'}$ Glu^{175} in KPI106. All structure models and *in silico* docking were performed by S. Heidt.

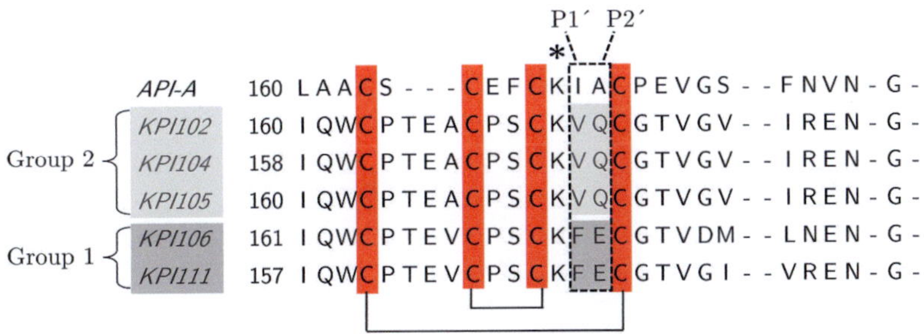

Figure 2.14: API-A P_1 Lys is conserved among the KPIs. Protein alignment highlighting the area around the API-A P_1 residue Lys[169] (marked with *), that is also conserved among the mycorrhizal-induced KPIs. According to their putative $P_{1'}$ and $P_{2'}$ residues (black dashed box), the mycorrhizal-induced KPIs were assorted into two groups. Group 1 is represented by KPI106 contains –FE– (dark gray box), whereas group 2, represented by KPI104, contains –VQ– in these positions (light gray box). The disulfide bond formation of the putative reactive site loop is indicated with black brackets and the respective cysteine residues are labeled in red.

contact to the catalytic Ser[226] of SCP1, suggesting that Lys[173] could be the $P_1$of KPI106. Comparison of the amino acid sequences of the mycorrhizal-induced KPIs showed that this Lys residue is conserved among them (Figure 2.14). A closer inspection of the residues next to this conserved Lys revealed that the KPIs contained different residues at the designated $P_{1'}$ and $P_{2'}$ positions. Based on that, the KPIs were categorized into two groups. Group 1, represented by KPI106, contains Phe[174]–Glu[175], whereas group 2, represented by KPI104, harbors the residues Val[171]–Gln[172] at the putative $P_{1'}$ and $P_{2'}$ positions. To experimentally assess the importance of the conserved Lys as well as its adjacent residues, mutant versions of KPI106 and KPI104 in which the respective residues were replaced by site-directed mutagenesis, were analyzed for interaction with SCP1.

In the first step, the conserved Lys was mutated to Gly in both inhibitors, and results showed that the interaction between SCP1 and KPI106[K173G] was stronger than with KPI106. Moreover, in case of KPI104, the Lys[170]Gly

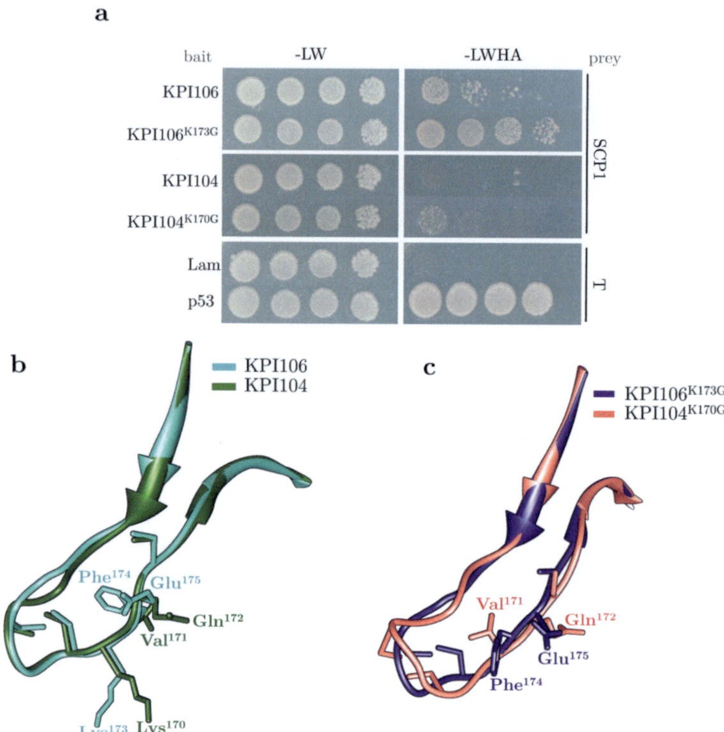

Figure 2.15: Mutation of the KPIs Lys$^{173/170}$ to Gly affects interaction affinity for SCP1. (a) The importance of the putative P_1 Lys$^{173/170}$ was assessed by mutation to Gly. KPI106^{Lys173Gly} showed a stronger interaction with SCP1 in a yeast two-hybrid test, compared to KPI106. Furthermore, KPI104^{Lys170Gly} did interact with SCP1 compared to KPI104. Controls: p53–T (positive), Lam–T (negative). **(b, c)** Three dimensional structure alignments of the loops containing Lys$^{173/170}$ of KPI106 (light green) with KPI104 (dark green, b) and KPI106^{Lys173Gly} (lilac) with KPI104^{Lys170Gly} (red, c) revealed that the Lys$^{173/170}$Gly mutation changes the steric conformation of the adjacent residues. Three dimensional alignments were performed by S. Heidt.

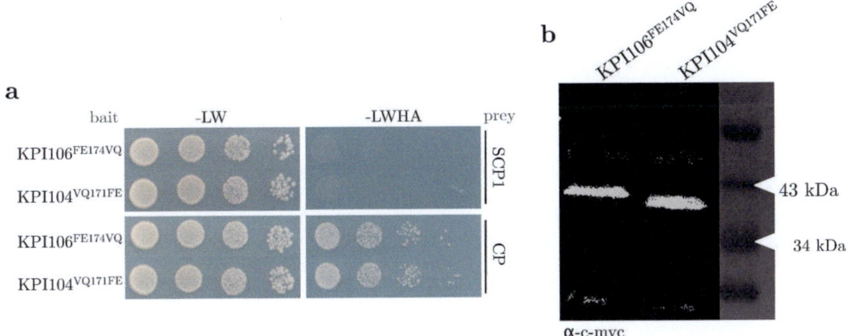

Figure 2.16: Putative $P_{1'}$ and $P_{2'}$ of KPI106 are necessary for the interaction with SCP1. **(a)** A chimeric version of KPI106, in which the putative $P_{1'}$ and $P_{2'}$ residues were switched to the according residues of KPI104 ($KPI106^{FE174VQ}$) is not able to interact with SCP1 anymore, as does the analogous chimeric inhibitor $KPI104^{VQ171FE}$. However, mutation of the putative $P_{1'}$ and $P_{2'}$ residues did not affect their interaction towards CP, indicating that Phe^{174}–Glu^{175} in KPI106 are specifically necessary to mediate the interaction with SCP1. **(b)** α-c-myc western analysis of yeast crude extracts confirmed the expression of the chimeric inhibitors $KPI106^{FE174VQ}$ (40 kDa) and $KPI104^{VQ171FE}$ (39 kDa).

mutation enabled an interaction with SCP1 (Figure 2.15a). Figure 2.15b–c shows three dimensional structure alignments of the KPI106 and KPI104 loops containing the native $Lys^{173/170}$ with the mutated versions containing $Gly^{173/170}$. It is apparent that the Lys to Gly mutation leads to a torsion within the loops that changes the steric conformation of the $Lys^{173/170}$ adjacent residues Phe^{174}–Glu^{175} in KPI106 and Val^{171}–Gln^{172} in KPI104.

Thus, in a second step, the importance of those residues was assessed. Therefore, chimeric versions of KPI106 and KPI104 in which the putative $P_{1'}$ and $P_{2'}$ residues were switched were constructed, resulting in $KPI106^{FE174VQ}$ and $KPI104^{VQ171FE}$. Subsequent yeast two-hybrid analysis revealed that the interaction between the chimeric $KPI106^{FE174VQ}$ and SCP1 was completely abolished, whereas the amino acid switch in $KPI104^{VQ171FE}$ was not sufficient to allow interaction with SCP1 (Figure 2.16a). Interestingly, both chimeric inhibitors were still able to interact with CP and the strength of these interactions did not differ from those obtained with the native inhibitors (see Figure 2.9c). In addition, to exclude false negative results in case of the abolished SCP1 interaction, the expression of both chimeric inhibitors was analyzed in yeast crude extracts. The western blot probed with α-c-myc in Figure 2.16b confirms that $KPI106^{FE174VQ}$ and $KPI104^{VQ171FE}$ were properly expressed, migrating at 40 kDa ($KPI106^{FE174VQ}$) and 39 kDa ($KPI104^{VQ171FE}$), respectively.

Taken together, these data indicate that Lys^{173} is likely the P_1 of KPI106. In fact, this residue is conserved among all mycorrhizal-induced KPIs and is important for the specificity and strength of the interaction between two individual KPIs and the protease SCP1. Moreover, the adjacent residues Phe^{174}–Glu^{175} in KPI106 are even necessary to mediate the interaction with SCP1, supporting the *in silico* docking model between the two proteins, and suggesting that SCP1 might be the target protease of KPI106 *in vivo*.

2.7 Subcellular localization of KPI106, KPI104, and SCP1

The SignalP4.1 and TargetP1.1 servers predicted that all KPIs as well as SCP1 contain an N-terminal secretion signal peptide, suggesting that the proteins might be targeted to the apoplast. In a first attempt, KPI106, KPI104, and SCP1 were expressed as C-terminal GFP fusions under control of the 35S promoter in *M. truncatula* hairy roots. However, confocal microscopy of the respective mycorrhizal roots revealed no distinct GFP signal, compared to the high autofluorescence of the hairy roots (see Figure S5). Thus, a second effort was made to verify secretion of those proteins in a heterologous system, and the same GFP fusion constructs

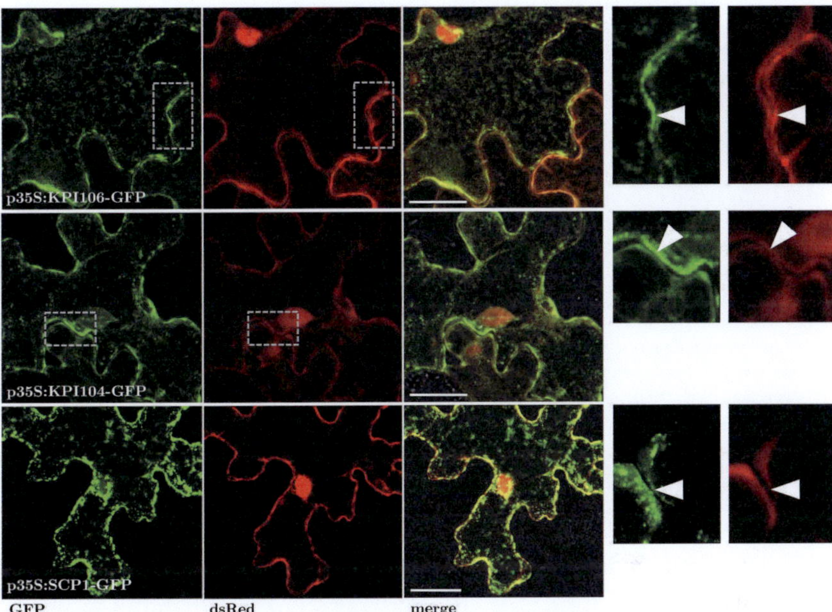

Figure 2.17: Localization of KPI106, KPI104 and SCP1. Heterologous expression of C-terminal GFP fusion proteins under the 35S promoter in *N. benthamiana*. GFP fluorescence of KPI106-GFP, KPI104-GFP, and SCP1-GFP was localized in dots around the cell periphery and the nuclei, indicating the secretory pathway of the fusion proteins. On the right, magnifications mark cell to cell border regions (white arrowhead). The magnification in case of SCP1-GFP is taken from a different image, as in the presented image on the left no adjacent cells were transformed. All images were created by maximum projections of z-stacks. White scale bar: 25 µm.

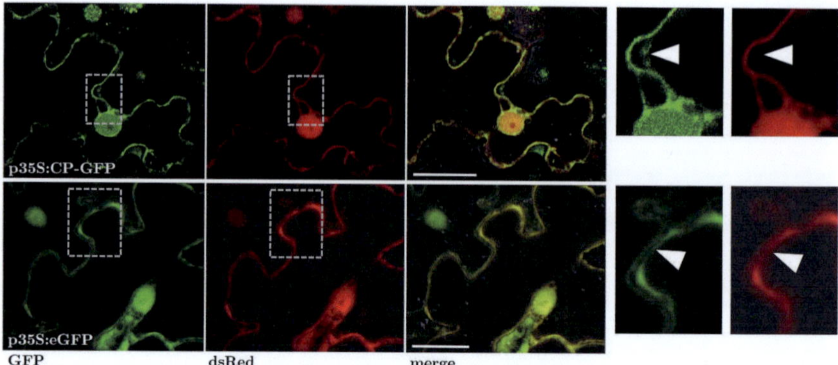

Figure 2.18: Localization controls. As control, CP-GFP that contains a vacuolar sorting motif showed a continuous distribution throughout the cell periphery and around the nucleus, whereas the weak GFP signal within the nucleus might be due to GFP cleave-off processes. Free DsRed marked the nucleus and cytoplasm, fully coincident with the localization of free GFP. On the right, magnifications mark cell to cell border regions (white arrowhead). All images were created by maximum projections of z-stacks. White scale bar: 25 µm.

were expressed in *N. benthamiana* epidermal cells. Free GFP as well as the previously identified CP, which was also predicted to contain a signal peptide but followed by the –NPIR– vacuolar sorting motif, were included as controls. Furthermore, DsRed under control of the ubiquitin promoter was used as control, to label the plant nucleus and cytosol.

Confocal microscopy revealed a similar localization pattern of both KPI-GFP proteins and SCP1-GFP. In all cases, the GFP fluorescence was observed in dots accumulating at the cell periphery as well as in the ER around the nucleus, indicating that the fusion proteins follow the conventional secretory pathway (Figure 2.17). Cell to cell borders were magnified for each construct, revealing that GFP was distinct from the DsRed fluorescence in these areas, extending further into the interspace between adjacent cells. In contrast, a dot-like localization was not observed in cells expressing CP-GFP (Figure 2.18). In this case, the GFP signal

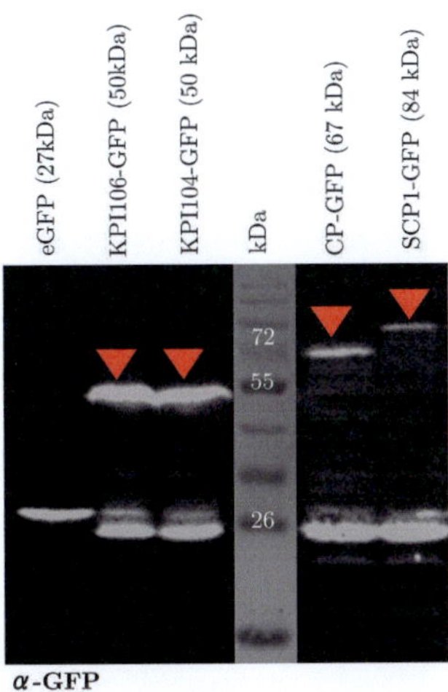

Figure 2.19: Confirmation of expression of heterologous fusion proteins in *N. benthamiana* leaf extracts. Integrity of C-terminal GFP fusions was analyzed by western blot with an anti-GFP antibody. GFP fusions migrate at the correct size (red arrow heads). In all lanes, free GFP is migrating at 27 kDa due to GFP cleavage during protein extraction.

was more continuously distributed throughout the cell periphery as well as around the nucleus. The weak GFP signal within the nucleus might be derived by GFP cleave-off processes from the protease. Magnification of a border region showed a distinct GFP fluorescence (white arrowhead in Figure 2.18) that was not present in the according magnified DsRed image, which could indicate the vacuole in this cell corner. In leaves infiltrated with p35S:eGFP, the GFP fluorescence was restricted to the cytoplasm and the plant nucleus, fully coincident with the DsRed localization in the magnification (Figure 2.18). Furthermore, the integrity of the GFP fusion proteins was demonstrated by western blot using an α-GFP antibody on tobacco protein cell extracts and all proteins KPI106-GFP (50 kDa), KPI104-GFP (50 kDa), CP-GFP (67 kDa) and SCP1-GFP (84 kDa) migrated according to their expected molecular sizes (Figure 2.19). In all lanes, free GFP resulting from cleavage during protein extraction was also observed migrating at 27 kDa.

2.8 Mycorrhizal-induced *SCPs* in *M. truncatula*

To address the question about potential homologs of SCP1, a blast screen using the *SCP1* cDNA sequence against the *M. truncatula* genome (hapmap3.5) was carried out. Figure 2.20 presents the identified 20 *SCP* genes in a phylogenetic tree, in which the *S. cerevisiae* carboxypeptidase Y (CPY; accession yeast genome database: YMR297W; Wolf & Fink, 1975) was included as outgroup. Twelve of the identified *SCP* genes were predicted (MtGEA) to be mycorrhizal-induced. Those genes can be further grouped into six pairs, each composed of two gene copies located on chromosome 1 (named *SCP1–6*) and chromosome 3 (named *SCP1a–6a*), respectively. Furthermore, on both chromosomes, the *SCP* genes are localized in clusters that display identical copies. Each *SCP* open reading frame is composed of seven exons and, Pfam analysis revealed all deduced SCP proteins to contain an N-terminal secretion signal peptide and a S10 protease domain harboring the catalytic residues Ser^{226}, Asp^{414}, His^{466} (Figure 2.21).

It is known that some SCPs do not exhibit proteolytic activity but function as acyltransferases (Lehfeldt *et al.*, 2000; Stehle *et al.*, 2009). However, true serine carboxypeptidases contain the diagnostic pentapeptide –GESYA– around the catalytic serine residue, in contrast to –GDSYS– in acyltransferases (Stehle *et al.*, 2009). The amino acid alignment of Figure 2.22 shows that all mycorrhizal-induced SCPs contain the –GESYA– motif, indicating that they are possibly true serine carboxypeptidases.

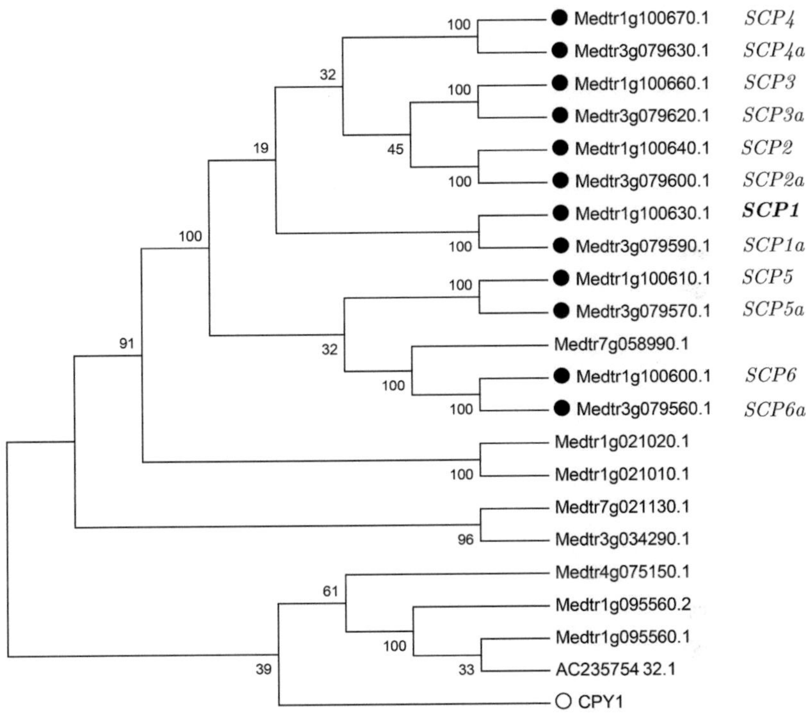

Figure 2.20: Homologs of *SCP1* in *M. truncatula*. 20 *SCP* genes were identified within the *M. truncatula* genome, whereas twelve of them (black dots) were predicted (MtGEA) to be exclusively induced during mycorrhizal symbiosis. The tree was constructed using the Neighbor-Joining algorithm and numbers are bootstrap supported values.

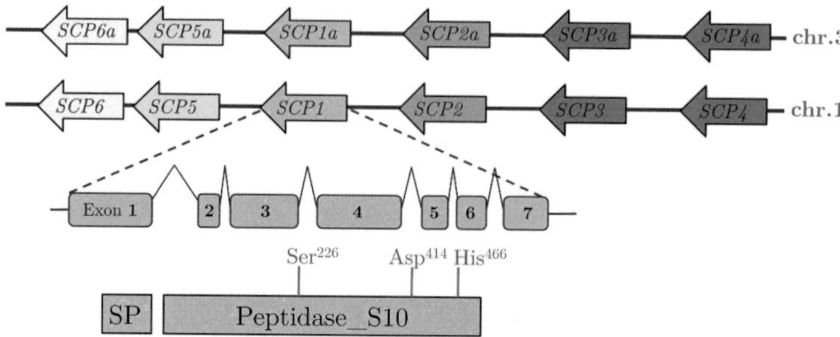

Figure 2.21: Genomic localization of mycorrhizal-induced *SCP*s. *SCP1–6* are localized on a cluster on chromosome 1, whereas a copy of this *SCP* cluster is present on chromosome 3, containing *SCP1a–SCP6a*. Each *SCP* gene is composed of seven exons, and proteins contain an N-terminal secretion signal peptide and a C-terminal S10 peptidase domain including the catalytic residues Ser^{226}, Asp^{414}, and His^{466}(amino acid numbers refer to the respective positions in SCP1).

```
                                                 *
SCP1   194 DKSTAKDAYVFLINWLERFPQYKTRDFYIT GESYA GHYVPQLASTILYNNKLYNNTI
SCP1a  194 DKSTAKDAYVFLINWLERFPQYKTRDFYIT GESYA GHYVPQLASTILYNNKLYNNTI
SCP2   192 DKSTAKDSYVFLINWLERFPQYKTRAFYIA GESYA GHYVPQLASTILHNNKLYNNTV
SCP2a  192 DKSTAKDSYVFLINWLERFPQYKTRAFYIA GESYA GHYVPQLASTILHNNKLYNNTV
SCP3   194 DKSTAKDAYVFLINWLERFPQYKTRDFYIT GESYA GHYVPQLASTILHHHKLYNKTI
SCP3a  194 DKSTAKDAYVFLINWLERFPQYKTRDFYIT GESYA GHYVPQLASTILHHHKLYNKTI
SCP4   194 DKSTAKDSYVFLINWLERFPQYKTRDFYIS GESYA GHYVPQLASTILHNNKLYKNTI
SCP4a  194 DKSTAKDSYVFLINWLERFPQYKTRDFYIS GESYA GHYVPQLASTILHNNKLYKNTI
SCP5   195 DKSTAKDAYVFLINWLERFPQYKTRAFYIT GESYA GHYVPQLASTILHNNKLYNNTT
SCP5a  195 DKSTAKDAYVFLINWLERFPQYKTRAFYIT GESYA GHYVPQLASTILHNNKLYNNTT
SCP6   194 DKSTAKDTYVFLVNWLERFPQYKTRDFYIT GESYA GHYVPQLASTILHNNKLYNNTI
SCP6a  194 DKSTAKDTYVFLVNWLERFPQYKTRDFYIT GESYA GHYVPQLASTILHNNKLYNNTI
```

Figure 2.22: SCPs are true serine carboxypeptidases. Amino acid alignment showing that all mycorrhizal-induced SCPs contain the highly conserved –GESYA– motif (gray box), which is characteristic for SCPs with proteolytic activity. Catalytic Ser is marked with (*).

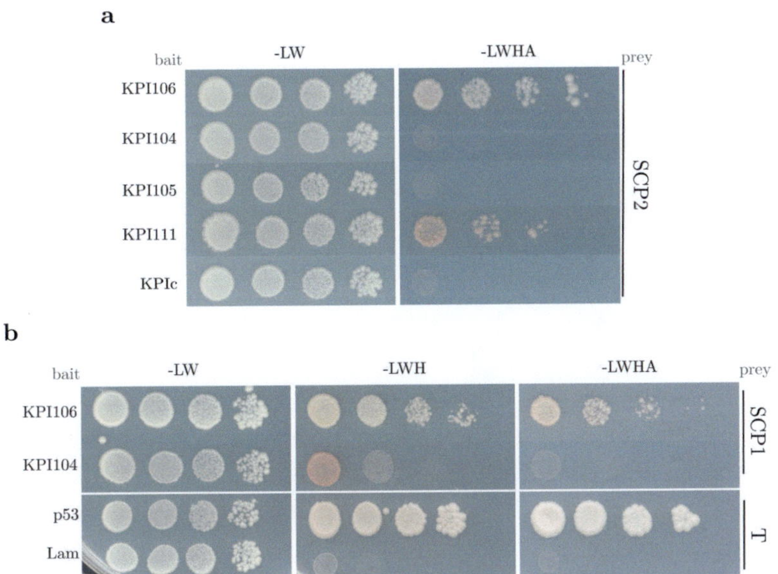

Figure 2.23: No differences of the KPIs interaction affinities for the SCP1 homolog SCP2. (a) Interaction of SCP2 with the mycorrhizal-induced Kunitz inhibitors revealed no differences in interaction affinities compared to their interaction with SCP1 (see Figure 2.12). Controls: p53–T (positive), Lam–T (negative), and KPIc–SCP2 (prey interaction with control inhibitor). **(b)** On the low-stringency -LWH medium, the yeast strain co-expressing KPI104 and SCP1 was able to grow, in contrast to the negative control Lam–T; thus, it cannot be excluded that the two proteins might interact *in vivo*.

As shown by the phylogenetic tree in Figure 2.20a, the predicted mycorrhizal-induced *SCP* genes share a very high sequence homology, suggesting that the deduced proteases might possibly have redundant or complementary functions. To assess this hypothesis, another member of the SCP cluster, SCP2, was analyzed for interaction with the mycorrhizal-induced KPIs in the yeast two-hybrid system. It has to be mentioned here, that compared to the other protease prey constructs, SCP2prey does not include the N-terminal prodomain. Results showed that, while the group

1 inhibitors KPI106 and KPI111 were able to interact with SCP2, no interactions were observed in case of KPI104 and KPI105 (Figure 2.23a). These results indicate that the prodomain is not necessary for the SCP2 interaction with KPI106 and KPI111, and reveal furthermore, that there are no differences compared to the interaction affinities as observed for SCP1 (see Figure 2.12). This could either indicate that the SCP proteins are not targets of the group 2 inhibitors KPI104 and KPI105 or that the interaction is too weak to be detected on the high-stringency -LWHA selection medium. Therefore, the yeast-interactions of KPI106 and KPI104 with SCP1 were analyzed on the low-stringency -LWH medium that contains additional adenine. In this case, the yeast strain co-expressing KPI104 and SCP1 was able to grow, even though the pinkish color indicated an adenine deficiency, but the overall growth of the co-transformant is enhanced compared to the negative control Lam–T (Figure 2.23b). Consequently, it cannot be excluded that KPI104 and SCP1 interact *in vivo*.

2.9 Inactivation of SCPs phenocopies the overexpression of KPI106 and KPI104

If individual members of the mycorrhizal-induced SCP family were among the protease target(s) of KPI106 and possibly KPI104 *in vivo*, it is very likely that a SCP loss-of-function could produce a similar phenotype as the overexpression of the KPIs. To test this hypothesis, the cellular function of the SCP family was examined in more detail.

First of all, the expression of the *SCPs* during mycorrhizal symbiosis was analyzed by real-time PCR. To monitor the entirety of the *SCP* transcripts, primer sequences were located in an area with high sequence similarity, and gene expression was analyzed in cDNA samples used for quantification of the *KPIs* transcripts (see Figure 2.2). Results showed that *SCP* transcripts are highly induced at 5 dpi and slightly increased up to the 17 dpi time point, congruent with the expression pattern of the mycorrhizal-induced KPIs (compare Figure 2.2a).

Secondly, a silencing approach was carried out by means of RNAi. Given the high nucleotide identity among the different *SCP* homologs, it was assumed that off-target silencing would occur. For better illustration purposes, a cDNA sequence alignment containing all *SCP* homologs and highlighting the area spanning the RNAi target as well as primer sequences used for qPCR can be reviewed in Figure S4. The successful silencing of all *SCP* homologs was confirmed by real-time PCR when comparing *SCP* expression in mycorrhizal *SCP*-silenced roots (17 dpi) and mycorrhizal control roots (Figure 2.24b). A morphological examination at 17 dpi revealed that *R. irregularis* colonized *SCP*-silenced roots contained numerous malformed arbuscules. Moreover, septation of intraradical fungal hyphae was frequently observed, in contrast to the empty vector control (Figure 2.24c–d). Interestingly, and as observed in KPI106 and KPI104 overexpressing roots, no changes in the frequency or intensity of fungal colonization occurred, whereas the overall number of mature arbuscules was significantly reduced in *SCP*-silenced roots, when compared to the empty vector control roots (red column, Figure 2.24e). Similarly, the mycorrhizal markers *PT4* and *RiMST2* were normally expressed in *SCP*-silenced roots and transcript abundance of *RiTEF* was comparable to that observed in the empty vector control roots (Figure 2.25; see Figure 2.11a for mycorrhizal markers in EV control roots).

Thus, the arbuscule phenotype of *SCP*-silencing indicates a crucial role of the SCP proteins for arbuscule development. Most importantly, this phenotype highly resembles the overexpression of KPI106 and KPI104 and suggests that KPI106 and possibly KPI104 together with SCP1 constitute tandems of which their functional interaction is involved in the control of arbuscule development within the root cortex of *M. truncatula*.

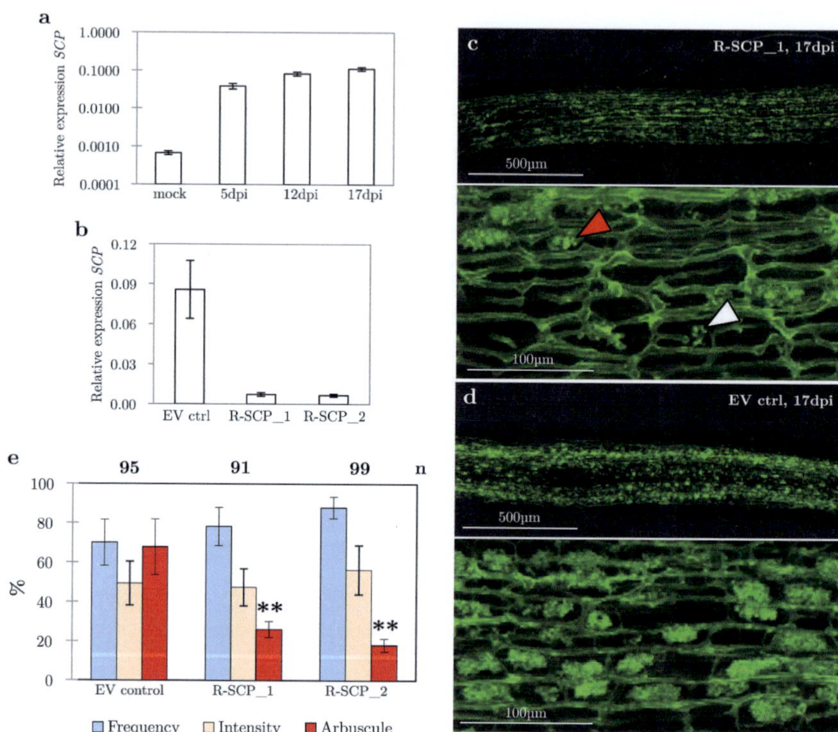

Figure 2.24: Silencing of SCPs leads to a similar phenotype as the overexpression of the KPIs. **(a)** Time course expression analysis of *SCP* during mycorrhizal symbiosis. *SCP* is highly induced at 5 dpi and slightly increased up to the 17 dpi time point. **(b)** Confirmation of successful silencing of *SCPs* in two RNAi lines by qPCR, 17 dpi. Values were normalized *TEF*. Error bars are standard deviations of means of three independent biological replicates. Visualization of *R. irregularis* fungal structures within *SCP*-silenced **(c)** and control lines **(d)** at 17 dpi, using WGA-Fluorescein. Many arbuscules in R-SCP_1 roots showed an aberrant morphology (red arrowhead) and intraradical hyphae contained septa (white arrowhead). Morphology of R-SCP_2 did not differ from that presented in R-SCP_1. **(e)** The frequency (blue column) and intensity (orange column) of mycorrhizal colonization were comparable in *SCP*-silenced and control roots, whereas arbuscule abundance (red column) was significantly lower in *SCP*-silenced roots, similar to the disturbed mycorrhizal morphology in case of KPI106 and KPI104 overexpression. **: p-value < 0.01, n: number of root fragments.

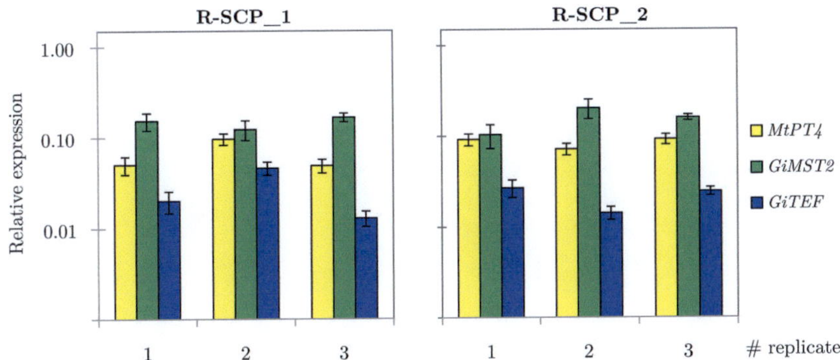

Figure 2.25: Expression of mycorrhizal markers in *SCP*-silenced roots. Expression levels of *M. truncatula PT4* and *R. irregularis MST2* and *TEF* were analyzed by qPCR in R-SCP and control lines at 17 dpi. Both genes are appropriately expressed in all tested samples, mirroring the results of marker gene expression in KPI106 and KPI104 overexpression lines. Transcripts were analyzed in colonized roots of three independent biological replicates (indicated with #) of each line. Errors represent standard deviations of three technical replicates and bars show relative expression values normalized to *MtTEF*, or *RiTEF* for *RiMST2*.

2.10 Perspective: Target identification of SCP1

To elucidate the biological role of SCP1 and its homologs, it is necessary to identify their protein targets. Given that many proteases are integral components of whole signaling networks, this is not a trivial task. Besides the recently emerging degradomics technologies (for example: Huesgen & Overall (2012)), a very basic approach to get a hint for protease substrates is a modified yeast two-hybrid system, called the "inactivated domain capture" that was developed by Overall *et al.* (2004). In this system, a mutant protease bait with an inactive catalytic domain is used to capture potential substrates within a yeast library, so that the protease is able bind with, but cannot cleave the substrate.

In like manner, a mutant version of SCP1, in which the catalytic Ser^{226} was replaced with Ala was constructed. However, the $SCP1^{S226A}$ bait

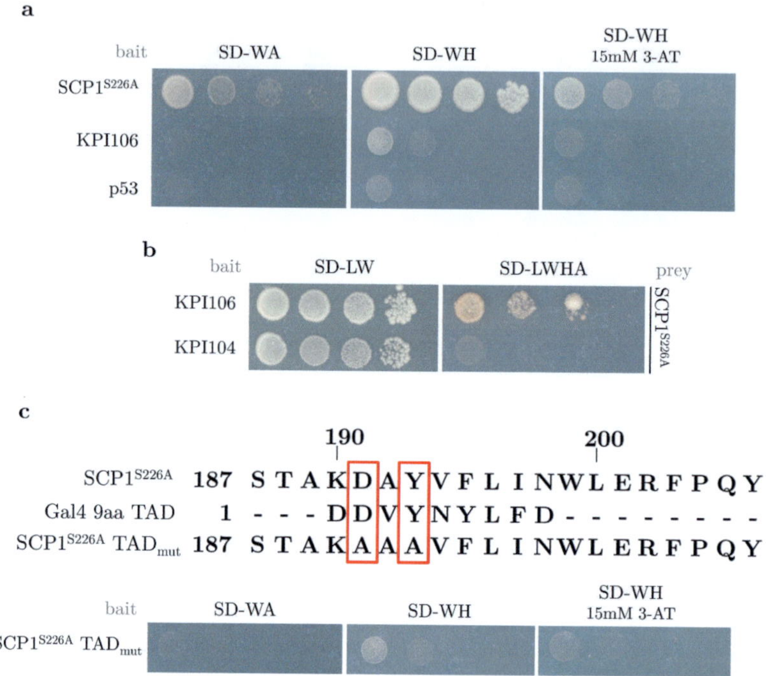

Figure 2.26: Construction of a yeast two-hybrid SCP1 bait protein. (a) $SCP1^{S226A}$bait shows high self-activity of the Ade2 and His3 reporters, when compared to KPI106 and p53 control baits. The $SCP1^{S226A}$ strain was able to even grow in presence of the competitive His3 inhibitor 3-AT [15 mM]. **(b)** The Ser^{226}Ala mutation was introduced into SCP1prey and the interactions KPI106–$SCP1^{S226A}$prey and KPI104–$SCP1^{S226A}$prey did not differ from those of native SCP1prey, indicating that Ser^{226}Ala was not responsible for self-activity. **(c)** Alignment showing the homology between SCP1 amino acids 190–198 and the 9aa TAD of the yeast Gal4 transcription factor. Mutations of the residues Asp^{191} and $Tyr^{193}$to Ala, resulting in $SCP1^{S226A}TAD_{mut}$ are highlighed in a red frame. $SCP1^{S226A}TAD_{mut}$ was not able to grow on -WA, -WH or -WH supplemented with 15mM 3-AT, confirming that self-activity was caused by the 190–198 aa fragment.

expressing yeast strain showed high self-activity of the Ade2 and His3 nutritional reporter genes, when compared to the KPI106- as well as the p53- control baits (Figure 2.26a). It is known that in transformants derived from the AH109 strain, the His3 reporter can be leaky due to specific bait properties (Fields, 1993). For that, the growth ability of the SCP1^{S226A}bait expressing yeast strain was tested in the presence of different concentrations of 3-AT, a competitive inhibitor of the yeast His3 protein. Results showed that growth on plates supplemented with 15 mM 3-AT was indeed reduced but still significantly stronger than growth of the control baits (Figure 2.26a). Using a bait with such high self-activity in a yeast two-hybrid assay would bias the screening towards false positives; thus SCP1^{S226A} is not suitable for a yeast two-hybrid screen. To exclude that self-activity is due to Ser^{226}Ala, this mutation was also introduced in the prey version of SCP1. Interaction affinities of KPI106–SCP1^{S226A} and KPI104–SCP1^{S226A} did not differ from those observed with the native SCP1prey protein (Figure 2.26b). This result suggests that SCP1^{S226A}bait possesses internal transactivating properties, since it only shows self-activity when expressed as bait fused to the DNA-binding domain of the Gal4 transcription factor, but not when expressed as prey fused to the transactivation domain of Gal4 (see Figure 2.12).

Eukaryotic transactivation domains are often characterized by a distinct motif of nine amino acids (9aa TAD; Piskacek *et al.*, 2007). However, *in silico* analysis using the EMBnet 9aa TAD server[1] did not predict a reliable transactivation domain within the SCP1 sequence. Therefore, the 9aa TAD motifs of prominent transcription factors including Gal4 of *S. cerevisiae* were aligned with the SCP1 protein sequence. Results revealed that the area between amino acids 190–198 of SCP1 shows homology to the 9aa TAD of the yeast Gal4 transcription factor (Figure 2.26c). To test if the

[1] www.es.embnet.org/Services/EMBnetAT/htdoc/9aatad/

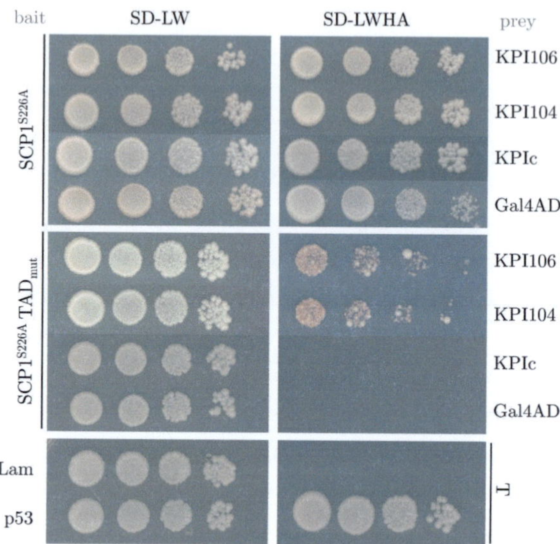

Figure 2.27: Interaction affinities of SCP1^{S226A}TAD$_{mut}$ bait. Yeast co-transformants expressing SCP1^{S226A}bait showed strong growth on high-stringency selection medium, whereas growth of SCP1^{S226A}TAD$_{mut}$–KPI106prey is weaker and thus comparable to the growth of KPI106–SCP1^{S226A}prey (Figure 2.26 b). Furthermore, yeast strains SCP1^{S226A}TAD$_{mut}$–KPIc and SCP1^{S226A}TAD$_{mut}$–Gal4AD did not grow, indicating that that the interaction between SCP1^{S226A}TAD$_{mut}$–KPI106prey is specific. However, SCP1^{S226A}TAD$_{mut}$–KPI104prey was able to grow on high-stringency selection medium, giving further proof that SCP1 could be a target protease of KPI104. Controls: p53–T (positive), Lam–T (negative).

190–198 aa fragment was responsible for SCP1^{S226A} self-activity, two of the matching residues, Asp^{191} and Tyr^{193}, were mutated to Ala, resulting in SCP1^{S226A} TAD$_{mut}$. Subsequent growth analysis on -WA, -WH and -WH supplemented with 15 mM 3-AT showed, that SCP1^{S226A} TAD$_{mut}$ was not able to self-activate the Ade2 and His3 nutritional reporters anymore, as growth was comparable to that of t the control baits KPI106 and p53. This suggests that the 190–198 aa caused the self-activity observed in the yeast two-hybrid assay (Figure 2.26c).

Next, to analyze the interaction properties of the new bait SCP1^{S226A} TAD$_{mut}$bait, it was co-transformed with either KPI106 as positive- as well as with KPI104, the control inhibitor KPIc, and the empty prey vector (Gal4AD) as negative prey controls, respectively. In addition, the interactions of all mentioned prey constructs with SCP1^{S226A} bait are presented in Figure 2.27 to illustrate the effect of Asp^{191}Ala and Tyr^{193}Ala mutations within the 190–198 aa fragment. Results showed that all co-transformants expressing SCP1^{S226A} bait had a strong growth on -LWHA, demonstrating its high self-activity. In contrast, a comparable growth of yeast strains co-expressing SCP1^{S226A} TAD$_{mut}$–KPI106prey and KPI106–SCP1^{S226A}prey (see Figure 2.26b) was observed. Furthermore, no growth was detectable in case of SCP1^{S226A}TAD$_{mut}$–KPIc or the prey vector control SCP1^{S226A} TAD$_{mut}$–Gal4AD, indicating that the interaction SCP1^{S226A}TAD$_{mut}$–KPI106prey is specific. Surprisingly, SCP1^{S226A}TAD$_{mut}$ did interact with KPI104prey and growth of this strain did not differ from that of SCP1^{S226A} TAD$_{mut}$–KPI106prey (Figure 2.27). This suggests that conformational issues when expressed as reverse bait and prey fusions might have diminished their interaction affinity, and might explain the observed growth of the co-transformant KPI104bait–SCP1prey on low stringency selection medium (compare Figure 2.23b).

In conclusion, these interaction studies showed that SCP1^{S226A} TAD$_{mut}$ could be an applicable bait protein for a future yeast two-hybrid screen and further support the previous observation that SCP1 could also be a likely target of KPI104 *in vivo*.

CHAPTER 3

DISCUSSION

3.1 Induction of Kunitz inhibitors during plant–microbe interactions

A family of Kunitz protease inhibitors is transcriptionally induced during mycorrhizal symbiosis in *M. truncatula.* Likewise, expression of the Kunitz inhibitor *SrPI1,* also part of a small inhibitor family, is enhanced in the legume *Sesbania rostrata* during nodule symbiosis with *Azorhizobium caulinodans* (Lievens *et al.*, 2004). Similar to the mycorrhizal KPIs, *SrPI1* expression is dependent on secreted molecules of the rhizobial symbiont, as bacterial strains defective of Nod factors were unable to induce *SrPI1* (Lievens *et al.*, 2004). In addition to their induction upon perception of diffusible molecules of *R. irregularis* (Kuhn *et al.*, 2010), various transcriptomic analyses report *KPI106* and *KPI104* to be up-regulated in response to other AM fungi including *Glomus mossae*, *G. versiforme* and

Gigaspora gigantea (Grunwald *et al.*, 2004; Hohnjec *et al.*, 2005; Liu *et al.*, 2007), suggesting that their induction is dependent on universal AM fungal molecule(s), but not on chitin (Kuhn, 2011, p. 67). In contrast, *KPI106* and *KPI104* are not induced by infection with *M. truncatula*-specific pathogens including *Colletotrichum trifolii* (Kuhn, 2011; p. 65–66) or the oomycete *Aphanomyces euteiches* (Nyamsuren *et al.*, 2007), suggesting that their function is not associated with a general pathogen-related defense response. This is in line with *SrPI1* expression that is not up-regulated upon challenge with host-specific pathogens (Lievens *et al.*, 2004). However, in the *M. truncatula*–*A. euteiches* pathosystem, two Kunitz inhibitors are induced in the infected root proteome (Schenkluhn *et al.*, 2010). Furthermore, silencing of one of them, *MtTi2* (matches the sequence of *Medtr6g078140.1* in Figure 2.1) influences the transcriptional expression of several plant genes in *A. euteiches*-infected *M. truncatula* roots (Nyamsuren *et al.*, 2007) This suggests that *M. truncatula* has at least two different sets of KPIs, one that is pathogen-inducible and functionally related to plant defense, whereas the other one, including *KPI106* and *KPI104*, is specific for the symbiotic interaction with arbuscular mycorrhizal fungi. In fact, there must be at least a third set of KPIs involved in rhizobial interactions.

3.1.1 Spatial expression of mycorrhizal-induced KPIs

Having discussed the temporal induction, the localization of KPI106-GFP and KPI104-GFP fusion proteins was assessed in a heterologous system, and results indicate their secretion as predicted by the N-terminal secretion signal peptide. Furthermore, the same GFP-fusion constructs were expressed in *M. truncatula* hairy roots, but confocal microscopy of those roots revealed no distinct GFP signal, compared to the high autofluorescence of the hairy roots. A defect within the expression constructs themselves can likely be excluded, as they are properly expressed in *N. benthamiana*.

Correspondingly, by analyzing the expression of p35:PT4-GFP, Pumplin *et al.* (2012) only detected a weak GFP signal similar to the intensity of the autofluorescing root vasculature. For this reason, the authors proposed that p35S is less active in *M. truncatula* cortical cells. This could likely explain the weak GFP signal within the p35S:KPI-GFP hairy roots as, in contrast, pUbi:DsRed showed a strong signal within the nuclei and at the cell periphery.

Formerly, Grunwald (2004) and Grunwald *et al.* (2004) showed in promoter-GUS experiments that a KPI encoding gene on TC69297, which matches the sequence of *KPI106* (MtGI), is expressed in arbuscule containing cortical cells of mycorrhizal *M. truncatula* roots (Grunwald, 2004; p. 82). In contrast, Kuhn (2011) analyzed 1 kB and 4 kB fragments of the *KPI106* promoter and results revealed in all cases a strong expression of the GUS reporter in mycorrhizal as well as in non-mycorrhizal control roots, indicating that the *KPI106* promoter is deregulated without its genomic context (Kuhn, 2011; p. 70). In conclusion, the applied methods to examine the localization of the KPIs indicate their secretion, but provide not enough information to give a detailed view of their spatial expression *in vivo*. Thus, prospective approaches to analyze their spatial expression during proceeding symbiosis would be either live cell imaging by means of a different constitutive promoter, as for example shown for the polyubiquitin promoter in *L. japonicus* (Maekawa *et al.*, 2008), or *in situ* hybridization as this method is independent of regulatory elements of the native promoter.

3.1.2 Mycorrhizal-induced KPIs contain six conserved cysteine residues

The mycorrhizal-induced KPIs possess six conserved cysteine residues. This is in contrast to most characterized KPIs of legume seeds that contain four cysteine residues forming two disulfide bonds (Oliva *et al.*, 2010).

Even though there are some individual seed-located KPIs, that differ from this default cysteine pattern, none of them harbors the triple disulfide bond pattern as indicated by the *in silico* model of KPI106 (Oliva *et al.*, 2010). In seeds, Kunitz inhibitors mainly function as protein storage for germination and the early seedling (Wilson *et al.*, 1988) or in combat of herbivory (Srinivasan *et al.*, 2005; Oliva *et al.*, 2011). In contrast, KPI106 and homologs are expressed in mycorrhizal roots, and the results of this work indicate that they could be involved in control of symbiotic fungal development by interaction with serine carboxypeptidases. It is therefore tempting to speculate whether the number of disulfide bonds in Kunitz inhibitors is in relation to their actual biochemical function (inhibition of proteases or as amino acid storage), or rather displays a structural adaptation to the physiological conditions of their working environment (mycorrhizal-induced KPIs are likely secreted into the apoplast).

KPI106 and homologs show structure and sequence similarities to the protease inhibitors API-A and API-B from *S. sagittifolia* (Bao *et al.*, 2009). In particular, all inhibitors contain six conserved cysteine residues, three disulfide bonds (predicted by the KPI106 *in silico* model), and a conserved Lys (Lys^{173} in KPI106) which could possibly represent their reactive site P_1 residue. In contrast to most described Kunitz inhibitors, API-A and API-B display double-headed protease inhibitors that contain two reactive sites and are able to bind two trypsin molecules simultaneously (Bao *et al.*, 2009). A closer inspection of the alignment of Figure 2.3a reveals that the second P_1 residue Leu^{111} of API-A is not conserved among the mycorrhizal-induced KPIs, suggesting that they do not posses an analogous second reactive site.

Interestingly, the conservation of six cysteine residues is quite common among plant KPIs in general. Out of the 50 closest homologs to the KPI106 sequence, 48 of them contained six conserved cysteine residues. Among them, the largest identified group derived from poplar, the model plant for woody trees (Bradshaw *et al.*, 2000). The poplar Kunitz inhibitor family is induced in leaves upon wounding or by herbivore challenge and the individual members exhibit a biochemical diversity in efficacy against different proteases. The authors propose a functional specialization among the different KPI members in combat of a multiplicity of pests, important for the maintenance of those long-lived plants (Major & Constabel, 2008). Thereby, Major & Constabel (2008) highlighted an interesting feature. By *in silico* prediction of three dimensional structures, they found that some of those KPIs contained six conserved cysteines but only two disulfide bonds. They explain these findings with dimer formation of two inhibitor molecules connected between the two surface exposed cysteines. In contrast, dimer formation of the mycorrhizal-induced KPIs can likely be excluded by means of the *in silico* model of KPI106 (Figure 2.13b).

3.2 Identification of putative target proteases

3.2.1 *M. truncatula* cysteine protease (CP)

CP interacts with KPI106 and KPI104 in yeast which is further confirmed by an *in vitro* pulldown. However, due to the different predicted subcellular destinations—KPIs are assumed to be secreted into the apoplast, whereas CP is predicted to be targeted to the vacuole (Ahmed *et al.*, 2000)—CP is unlikely a target of the KPIs *in vivo*. This is in line with the highly complementary interaction between KPI106 and SCP1 in the docking model, suggesting that KPI106 is rather an inhibitor of serine than of cysteine proteases. Even though there is evidence for Kunitz protease inhibitors with a double activity towards serine and cysteine proteases (Migliolo *et al.*, 2010), the majority of the identified KPIs inhibits just one type of protease (Rawlings *et al.*, 2004). Additionally, the different phenotypes produced by silencing of CP and overexpression of KPI106 and KPI104 further support the hypothesis that CP is not their target protease *in vivo*.

CP belongs to the class of papain-like C1A proteases that were reported to be involved in many general physiological processes, including senescence and programmed cell death. Furthermore, secreted papain-like CPs play an important role in plant–pathogen associations (van der Hoorn, 2008; Martínez *et al.*, 2012). In line with the prediction by the MtGEA, *CP* is not among the mycorrhizal-induced *M. truncatula* cysteine proteases (Liu *et al.*, 2003; Hohnjec *et al.*, 2005). However, as a matter of fact, RNAi-mediated silencing of *CP* leads to an arbuscule phenotype. But at the same time, CP-silenced roots show unspecific symptoms including a brownish coloring and a bushy growth. This indicates that the function of CP is somehow required for arbuscule development. Thus it is not clear whether this is a direct involvement, indicative of a specific phenotype, or indirect, implying that the phenotype is caused due to general impairments of the CP-silenced roots. The fact that cysteine proteases are generally involved in physiological processes, together with the mycorrhizal-independent expression of *CP*, suggest that the CP-silenced phenotype is rather unspecific.

However, the activity of proteases is not only controlled on the transcriptional level, as shown by the example of the vacuolar-associated cysteine protease PsCYP15A of pea (Vincent & Brewin, 2000). In particular, western blot analyses showed that the inactive pro- (~38 kDa) and mature forms (~30 kDa) of PsCYP15A were generally present in most plant tissues, but when pea is engaged in root nodule symbiosis, only the mature form was found in isolated symbiosomes. Therefore, the authors propose a dual function of PsCYP15A. On the one hand, PsCYP15A would fulfill a housekeeping function related to the general metabolism and secondly, the cysteine protease would contain a specific function in root nodule symbiosis, probably involved in protein turnover within the symbiosome compartment

(Vincent & Brewin, 2000). Moreover, when the *M. truncatula* homolog of PsCYP15A was silenced by nodule-specific expression of an RNAi-target, the senescence of nodules was delayed (Sheokand *et al.*, 2005; Vorster *et al.*, 2013).

Likewise, a dual function of CP including an additional, mycorrhiza-related activity would be conceivable. CP is predicted to be transported to the vacuole, which, like all organelles of cortical cells, has to be reorganized in order to accommodate the entering fungus and the developing arbuscule (Cox & Sanders, 1974; Scannerini & Bonfante-Fasolo, 1983). In particular, it is assumed that the vacuole is partially fragmented in arbusculated inner cortical cells (Pumplin & Harrison, 2009). In this context, a specific arbuscule phenotype of the CP-silenced roots could be explained by the lack of a CP-processed signal that is required for subcellular reorganization of the inner cortical cells to harbor the developing arbuscule. If so, a spatio-temporal antisense inhibition (Sheokand *et al.*, 2005) of CP under control of an arbuscule exclusive promoter would reproduce the arbuscule phenotype, but not the unspecific symptoms of CP-silenced roots.

3.2.2 Mycorrhizal-induced *M. truncatula* proteases

Interaction tests in yeast between SCP1, the *M. truncatula* closest homolog of LjSbtM1 (Takeda *et al.*, 2009), and the mycorrhizal-induced Kunitz inhibitors revealed that SbtM1 is probably not a target protease of the KPIs. In contrast, the strong and specific interactions between KPI106–SCP1, KPI111–SCP1 and most likely $SCP1^{S226A}TAD_{mut}$–KPI104 indicate that SCP1 could be the target protease of those inhibitors. This is in line with API-A that is also an inhibitor of serine proteases (Bao *et al.*, 2009). Correspondingly, a distinct loop of KPI106 perfectly fits into the active pocket of SCP1, indicating that KPI106 follows the conventional canonical standard mechanism of protease interaction (Laskowski, 1968; Laskowski

& Qasim, 2000) and suggesting that this loop represents the main SCP1-binding domain of KPI106.

The experimentally assigned P_1 residue of API-A is Lys^{169} (Bao *et al.*, 2009) and mutation of the analogous Lys^{173} in KPI106 results in a stronger interaction with SCP1. Given that substitutions of the P_1 residue in protease inhibitors affect their target protease specificity (Laskowski & Kato, 1980), these results strongly suggest that this conserved Lys is the P_1 of the mycorrhizal KPIs. In case of KPI104, the Lys^{170}Gly mutation has an even more distinct effect, as it enabled an interaction with SCP1. A likely explanation could be that the switch from the bulky Lys to the small Gly induced more flexibility into the respective KPI loops—explicit by their three dimensional structure alignments—which, in turn, leads to a higher affinity for SCP1. With respect to the default positive interaction between $SCP1^{226A}TAD_{mut}$bait–KPI104prey when swapping bait and prey, the Lys^{170} Gly mutation probably led to overcome conformational issues that blocked the interaction between KPI104bait–SCP1prey, as the use of artificial fusion proteins in the yeast two-hybrid system bears a potential risk for misconformation associated with the inaccessibility of binding sites (Criekinge & Beyaert, 1999).

In protease–protein inhibitor interactions that follow the canonical standard mechanism, next to the contacts of the P_1 residue, the contacts between the adjacent P_2, $P_{1'}$, and $P_{2'}$ residues and their opposing protease subsites are important for mediating protease specificity (Bode & Huber, 1992). In line with this, three dimensional structure alignments show that the $Lys^{173/170}$Gly mutations in KPI106 and KPI104 lead to a torsion of the Lys adjacent residues within this exposed loop. Moreover, a switch of the designated $P_{1'}$ and $P_{2'}$ residues in KPI106 ($Phe^{174}$and Glu^{175}) with the according residues of KPI104 (Val^{171} and Gln^{172}) completely abolished the interaction with SCP1, which is consistent with other studies showing that replacements of P_1 adjacent residues lead to alterations in protease binding properties and/or specificity of the inhibitor (for example: Salameh *et al.*, 2010; Navaneetham *et al.*, 2013). Mueller *et al.* (2013) showed that amino acid deletions or mutations within the functional inhibitor domain of the *U. maydis* protease inhibitor Pit2 result in a loss of the physical interaction with its target proteases in yeast. In case of KPI106, the putative $P_{1'}$ Phe^{174} and $P_{2'}$ Glu^{175} residues are necessary to mediate the interaction with SCP1. In contrast, $KPI106^{FE174VQ}$ and $KPI104^{VQ171FE}$ are still able to interact with CP, suggesting that the KPI–CP interaction is not depending on the putative reactive site loop of the KPIs.

In summary, these results suggest that Lys^{173}, Phe^{174}, and Glu^{175} represent the respective P_1, $P_{1'}$ and $P_{2'}$ residues of KPI106 as they are part of a distinct loop that is in close contact to the predicted catalytic triad of SCP1. Mutational analyses confirm their importance in mediating strength and specificity of the interaction with SCP1, as substitutions of random KPI106 residues that are not part of the interacting domain might not produce such severe effects like a loss of physical interaction with SCP1. Coupled with the characteristic sequence feature of the conserved cysteine residues, these data indicate that at least KPI106 is a true Kunitz inhibitor since it interacts with a serine protease by blockage of the active site with a canonical loop. This is default for the majority of the so far characterized Kunitz inhibitors (Rawlings *et al.*, 2004). Even though it is conclusive that protease inhibition is the consequent result of this interaction, these data do not fully prove the inhibitory activity of KPI106. However, the fact that the overexpression of KPI106 and KPI104 as well as silencing of the SCP proteins produce highly similar phenotypes indicates that both types of proteins are involved in control of the same process *in vivo*. This is further supported by the comparable localization of KPI106 and SCP1 in *N. benthamiana* epidermal cells and their similar transcriptional expression patterns during proceeding mycorrhizal symbiosis.

3.3 Analysis of the KPI overexpression and SCP-silenced phenotype

Overexpression of KPI106 and KPI104 under the 35S promoter causes a mycorrhizal phenotype consisting of many malformed arbuscules, a reduced number of maturely developed arbuscules, and septa formation within intraradical hyphae; a phenotype also observed when silencing the SCP proteins. Together with the interaction analyses and the localization data presented in this work, the similarity of those phenotypes displays a further hint that the inhibitors and the serine carboxypeptidases control the same process *in vivo*. Therefore, both phenotypes shall be discussed together in the following section.

In order to integrate the malformed arbuscule morphology of the KPI overexpression and SCP-silenced phenotype into the current model of arbuscule development (Gutjahr & Parniske, 2013; see Figure 1.5), it is necessary to also take into account the applied experimental system that was used for the mycorrhizal colonization. In this work, the mycorrhization assays were performed using the two compartment plate system (Kuhn *et al.*, 2010; see Figure 5.1). In this system, the development of mycorrhizal symbiosis cannot be synchronized, and with proceeding colonization the full spectrum of the different arbuscule stages, from developing to collapsing, is present. In both mycorrhizal KPI overexpression as well as SCP-silenced roots, fully mature arbuscules could be observed at 17 dpi, even though less abundant. This suggests that the phenotype could be related to an early induced arbuscule turnover. However, the genetic marker *PT4*, which was shown to maintain arbuscule maturity (Javot *et al.*, 2007), is properly expressed in both colonized roots. This indicates that an early arbuscule turnover as observed for the *pt4* mutant (Javot *et al.*, 2007) is rather unlikely the cause for the KPI overexpression/SCP-silenced phenotype.

Arbuscule mutants can be divided into two groups. One includes mutants with reduced or absent *PT4* expression (for example: Floss *et al.*, 2008a; Takeda *et al.*, 2009; Baier *et al.*, 2010), whereas mutants of the second group reveal a malformed arbuscule morphology accompanied by *PT4* levels comparable to wildtype (Zhang *et al.*, 2010; Gutjahr *et al.*, 2012; Groth *et al.*, 2013). Consequently, the KPI overexpression/SCP-silenced phenotype must be assigned to the second group. Among those arbuscule mutants with unaltered *PT4* expression are the *str/str2* mutants. *STR/STR2* encode for two half-size ABC transporters and the two proteins specifically heterodimerize in the branching region of the peri-arbuscular membrane (Zhang *et al.*, 2010; Gutjahr *et al.*, 2012). Likewise, the *L. japonicus* mutant *SL0154* showed a malformed arbuscule phenotype in which only the arbuscule trunks as well as the initial branches were developed and *PT4* levels were comparable to wildtype (Groth *et al.*, 2013). In summary, the mycorrhizal phenotypes of *str/str2* and *SL0154* show a clear consistency to the phenotype of the KPI overexpression/SCP-silenced roots. This suggests that the KPI overexpression/SCP-silenced phenotype could be attributed to a developmental arrest of arbuscule formation in a low-order branching state. The observation of occasional mature arbuscules within the colonized KPI overexpressing/SCP-silenced roots could be assigned to leaky amounts of the *SCP* transcripts that were detected within both SCP-silenced roots, as well as that the amount of inhibitors within the KPI overexpression roots is not sufficient to block all molecules of their respective target protease.

In addition, the intraradical hyphae within the colonized KPI over-expressing and SCP-silenced roots contain noticeable amounts of septa. Arbuscular mycorrhizal fungi are coenocytic and the occurrence of septated hyphae is a signature of fungal decay (Bonfante-Fasolo, 1984). However, the fungal colonization of the KPI overexpression and SCP-silenced roots is comparable to the colonization of the empty vector control roots. This is in line with the similar expression levels of the fungal marker genes *RiTEF* and *RiMST2* in those roots; thus making it unlikely that active fungal proliferation within those roots is affected. While septa formation within arbuscules was attributed to arbuscule collapse (Bonfante-Fasolo *et al.*, 1990), the occurrence of septa within the intraradical hyphae is less common. However, pre-symbiotic AM fungal hyphae start septating if they cannot detect an appropriate plant host and subsequently retrieve the cytoplasm back into the spore (Logi *et al.*, 1998). In this context, the formation of septa within the intraradical hyphae of colonized KPI overexpression/SCP-silenced roots could be interpreted as an energy saving mode of the fungus, when detecting the arbuscule formation defect.

In contrast to the overexpression experiments, RNAi-mediated silencing of *KPI106* and *KPI104* did not produce a visible mycorrhizal phenotype, indicating that the KPI homologs act redundantly (Figure S2). This is in line with KPI106 and KPI111 both interacting with the same proteases SCP1 and SCP2, and suggests that the lack of one of them could be functionally compensated by the other. In fact, in KPI106 RNAi-lines, no changes in the expression of *KPI111* or the more distant related *KPI104* could be observed (see Figure S1), indicating that there was no cross-silencing involved. An option to investigate the KPI loss-of-function would be the use of a mutant in which several of the KPIs are affected. In this context, the recently released *M. truncatula* TILLING mutant collection of RevGenUK[1] shall be mentioned. TILLING (Targeting Induced Local Lesions IN Genomes; Colbert *et al.*, 2001) mutants are based on seeds treated with ethylmethanesulphonate to induce random point mutations. Having a mutant in which at least one of the KPIs is non-functional, a redundant inhibitor could be eliminated by RNAi-targeting.

[1] revgenuk.jic.ac.uk/about.htm

3.3.1 Possibility for tandem interactions between mycorrhizal SCPs and KPIs?

The results of this work suggest the possibility for putative tandem interactions between the Kunitz inhibitors and the serine carboxypeptidases that shall be discussed in the following section.

The *M. truncatula SCPs* share a high sequence identity, indicating that they may have redundant or complementary functions. They are exclusively induced during mycorrhizal symbiosis (MtGEA), which was experimentally confirmed in this work as they are fully induced at 5 dpi, whereas none of the transcripts was expressed in mock-treated roots when using primers that bind in a homologous region. Correspondingly, the KPIs are mycorrhiza-dependent expressed and KPI106, KPI104 and SCP1 show analogous localization patterns in *N. benthamiana*. Furthermore, the SCP proteins have a crucial role for the symbiosis as indicated by the arbuscule phenotype in SCP-silenced roots, that also occurs when overexpressing the Kunitz inhibitors KPI106 and KPI104. The interaction tests in yeast revealed KPI106 and KPI111 as specific counterparts of SCP1 and SCP2. Interestingly, those two inhibitors contain the residues –Phe-Glu– at their putative $P_{1'}$ and $P_{2'}$ positions (group 1 inhibitors) that are necessary to mediate the interaction between KPI106 and SCP1. In contrast, KPI105 and KPI104 (group 2 inhibitors) that harbor –Val-Gln– at the analogous positions, did not interact with SCP1 in the initial yeast tests. A closer inspection of the KPI encoding genes indicates that this target protease specificity might be mirrored by the localization of the respective genes. *KPI106* and *KPI111* are located consecutively on chromosome eight, whereas the group 2 genes *KPI105*, *KPI102* and possibly *KPI104* are located in a row on chromosome three. This finding could be interpreted as that the two groups of KPIs might have evolved from gene duplications and that selective pressure may have created a divergence towards their target protease specificities. Now, the question is whether the selective pressure on group 2 was to inhibit redundant versions of SCP1, which is already covered by functional redundancy among the group 1 inhibitors KPI106 and KPI111 that both interacted with SCP2. Furthermore, it is not clear yet whether KPI104 is a true interactor of SCP1 as they only interacted in the combination $SCP1^{S226A}TAD_{mut}$–KPI104prey, but not as KPI104bait–SCP1prey in the yeast two-hybrid system. This interaction needs to be further examined. In addition, not all SCP homologs have been tested for interaction with the KPIs and with regard to the similar arbuscule phenotypes of KPI104 overexpression and SCP-silencing it is tempting to speculate whether the target proteases of the group 2 inhibitors are among SCP3–6.

Ultimately, a definitive biochemical proof for the functional interaction between the SCP proteins and the KPIs is outstanding. The proper expression of SCP1 as C-terminal GFP fusion in *N. benthamiana* suggests that a heterologous expression in a plant system is favorable. A future approach could be the use of activity-based protease profiling (ABPP; van der Hoorn *et al.*, 2004) that has recently grown in popularity to characterize apoplastic plant proteases (for example: van der Linde *et al.*, 2012a; Lozano-Torres *et al.*, 2012; Mueller *et al.*, 2013). In this system, heterologously overexpressed plant proteases are isolated from apoplastic fluids of tobacco to avoid cleavage of the affinity tag. Those protease-containing apoplastic fluides are then co-incubated with the inhibitor proteins of interest as well as with artificial inhibitor molecules. This creates a competition between the putative inhibitors and the artificial molecules for binding to the active center of the protease. If it is possible to subsequently pull-down the protease by means of the affinity tag of the artificial molecules, then the tested inhibitor proteins were not able to outcompete the artificial molecules for interaction with the active center of the protease and thus did not inhibit the protease.

3.3.2 Towards the function of mycorrhizal-induced serine carboxypeptidases

The results of this work showed that members of a mycorrhizal-induced Kunitz inhibitor family interact with at least two serine carboxypeptidases that are also AM-specific. This raises the question what process requires the proteolytic activity of the SCPs?

Serine carboxypeptidases cleave off C-terminal amino acids from their target proteins. In mammals, many SCPs are involved in signaling events, for example to generate bioactive peptides in the blood coagulation cascade (Hagen *et al.*, 1996; Turk, 2006) or the biosynthesis of neurotransmitters and peptide hormones. In particular, the anterior pituitary gland hormone corticotropin is synthesized as a larger precursor and is post-translationally cleaved by an endopeptidase at basic amino acid containing sites (Hashimoto *et al.*, 1994). Subsequently, those exposed basic residues at the neo-C-termini are cleaved-off by SCPs, transforming the inactive intermediate into the active hormone (Fan *et al.*, 2002). Another function of SCPs is the post-translational processing of tubulins (Berezniuk *et al.*, 2012). This process is essential, as *pcd* mice in which the cytosolic carboxypeptidase 1 is mutated Fernandez-Gonzalez *et al.*, 2002 showed neuronal cell degeneration within a few weeks after birth (Mullen *et al.*, 1976).

In plants, recent studies suggest that SCPs might also be involved in signaling events important for plant growth and development. For example, SCPs were found to be induced by the plant hormone gibberellic acid in seeds of various plant species, where they participate in the mobilization of storage proteins for nutrition of the seedling (Bethke *et al.*, 1997; Bewley, 1997; Dominguez & Cejudo, 1999; Cercos *et al.*, 2003). In *Arabidopsis*, overexpression of the secreted serine carboxypeptidase BRS1 in a *bri1* mutant background is able to suppress the dwarfed phenotype

(Li *et al.*, 2001). *Bri1* mutants are affected in the receptor-like kinase BR (Clouse *et al.*, 1996). Interestingly, the serine carboxypeptidase BRS1 only complements the *bri1* phenotype of mutants that carry mutations in the extracellular domain of BR, whereas BRS1 fails to suppress the respective kinase dead mutant (Li *et al.*, 2001). BR contributes to the perception of the brassinoid growth hormone and it is proposed that BRS1 is involved in an early step of the BR signal transduction, possibly by processing a protein that may directly or indirectly participate in BR perception (Li *et al.*, 2001; Zhou & Li, 2005).

The *M. truncatula* serine carboxypeptidase SCP1 is transcriptionally induced in cells passed by the penetrating fungus, but also in adjacent cells that do not harbor any fungal structures (Liu *et al.*, 2003). It is thus proposed that the secreted serine carboxypeptidase SCP1 would produce a bioactive peptide signal that could either act cell-autonomous or cell-nonautonomous. This signal would be part of a plant program to control or predetermine fungal colonization within the root, as uncontrolled fungal spread might be harmful to the plant. In mycorrhizal symbiosis, the operating mode of cell-nonautonomous signals is evident, as in cortical cells that lie ahead of the fungal colonization front, PPA formation is induced in anticipation of fungal penetration (Genre *et al.*, 2005; Genre *et al.*, 2008). The root colonization by AM fungi culminates in formation of the highly branched arbuscules, which in turn, poses drastic subcellular reorganization on the inner cortical cells. Based on the arbuscule phenotype of SCP-silenced roots, the lack of the SCP-processed signal would make the plant to stop in promoting arbuscule development inside the inner cortical cells. This would result in a developmental arrest of the arbuscules in a low order branching state. As a consequence of this developmental block, the intraradical AM fungal hyphae start septating in order to no longer engage in an interaction that was already abolished by the plant. In this scenario, the secreted Kunitz inhibitors would provide a fine-tuned mechanism to regulate the spatial and temporal activity of the SCPs (Figure 3.1). As serine carboxypeptidases only cleave C-terminal residues, it could be possible that the SCPs are part of a cascade that, at the end, produces the bioactive signal. As observed for the processing of hormone precursors in mammals, it would be conceivable that an upstream acting endopeptidase cleaves a common target protein and thus generates neo-C-termini, that are subsequently trimmed by the SCPs to produce the active signal. In this context, it is worth to mention that the expression pattern of the mycorrhizal-induced subtilase SbtM1 is intriguingly similar to the pattern of SCP1.

3.4 Concluding remarks and outlook

Proteinaceous protease inhibitors control proteolytic cleavage that is required for a striking variety of physiological processes in all organisms. The fundamental role of secreted proteases and their inhibitors to promote compatibility or to mediate disease resistance in plant–pathogen associations was recently uncovered (Shindo & van der Hoorn, 2008; van der Linde *et al.*, 2012a). It is known that some molecular mechanisms, which are required to establish plant–pathogen relationships, are also involved to promote the symbiotic association between plants and AM fungi (Kloppholz *et al.*, 2011). This attempted to investigate the mycorrhizal-induced KPIs, as secreted protease inhibitors could possibly target endogenous, as well as fungal proteases. Thus, in frame of my previous Diploma thesis, a yeast two-hybrid screen using the KPI106 bait protein to identify putative targets within a cDNA library of *R. irregularis* colonized *M. truncatula* roots was employed. The results suggested that KPI106 likely targets endogenous proteins, as none of the identified positive clones were of fungal origin. However, it cannot be excluded that KPI106 also acts on

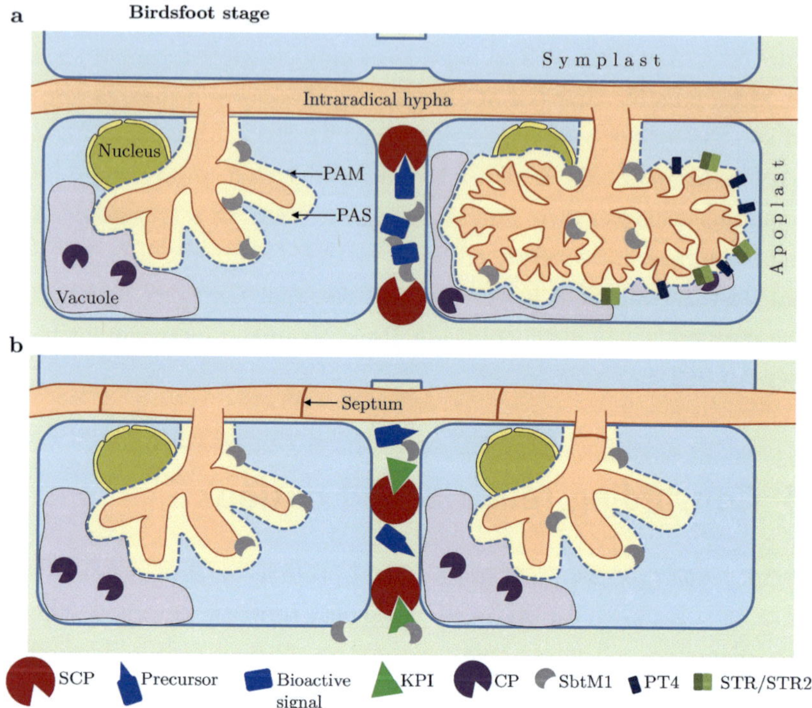

Figure 3.1: Summary of results and proposed model. **(a)** Schematic diagram of arbuscule development according to Gutjahr & Parniske (2013) showing the transition from the low-order branching state called "birdsfoot" to the mature arbuscule. Based on the results of this work it is proposed that within the apoplastic space, the SCPs would process a precursor protein in order to generate a bioactive signal which is part of a plant controlled program to coordinate arbuscular mycorrhizal development within the root cortex. In addition, the subtilase SbtM1 (Takeda *et al.*, 2009) that localizes to the apoplast as well as to the PAS, the PAM-resident heterodimer of the ABC half-size transporters STR/STR2 (Zhang *et al.*, 2010; Gutjahr *et al.*, 2012; respective *str/str2* mutants produce a similar arbuscule phenotype as observed for the KPI overexpressing and SCP-silenced roots), as well as the phosphate transporter PT4 (Harrison *et al.*, 2002; Javot *et al.*, 2007) are included in this scheme. The identified cysteine protease CP is predicted to be targeted to the vacuole and CP-silencing leads to a mycorrhizal phenotype. However the specificity of this phenotype needs to be further investigated. **(b)** SCP-silencing or overexpression of the KPIs would drastically reduce the amount of the processed bioactive signal which leads to a developmental arrest of the arbuscules in the birdsfoot state and consequently to the formation of septa within the intraradical hyphae.

fungal proteins *in vivo*. In this study, the functional characterization of the mycorrhizal-induced Kunitz protease inhibitors was continued, and results revealed that they most likely act as canonical inhibitors of several members of a family *M. truncatula* serine carboxypeptidases. The importance of both protein families for AM symbiosis is evident as deregulation of two of the Kunitz inhibitors, as well as silencing of the SCPs led to a highly similar arbuscule phenotype, supporting that members of the two families might functionally interact *in vivo*. Based on the results of this study, it is proposed that the Kunitz inhibitors control the spatial and temporal activity of the serine carboxypeptidases, and that the SCPs would produce a bioactive peptide signal as part of a plant program to control arbuscular mycorrhizal development within the root cortex. The most intriguing prospective task will be to identify the target protein(s) of the SCPs. Furthermore, additional information about the specific component/s or elicitor/s that induce the expression of the KPIs and SCPs, together with the knowledge of the SCP target protein(s) would enable to narrow their function within the physiological processes that control arbuscule development within the root cortex of *M. truncatula*.

CHAPTER 4 MATERIALS

4.1 Chemicals, kits, enzymes, buffers and solutions

4.1.1 Chemicals

All chemicals and consumable materials used in this study were of research grade and were obtained from AppliChem, Duchefa Biochemie, Fluka, Merck, Carl Roth, and Sigma Aldrich.

4.1.2 Kits and miscellaneous materials

The following kits were used as described by the supplier: TOPO TA Cloning® Kit and pENTR™/D-TOPO® Cloning Kit (Invitrogen) were used to directly clone PCR products. Gateway® LR Clonase® II Enzyme Mix (Invitrogen) was used for recombination of DNA fragments between

Table 4.1: Standard solutions used in this study.

Solution (conc.)	**Composition**
3-AT (1 M)	0.084 g/ml 3-AT in $_{d}H_2O$, sterile filter (0.22 µm), light protected storage at 4°C
Chemiluminescent A	0.25 mg/ml luminol in 0.1 M Tris-HCl pH 6.8, light protected storage at 4°C
Chemiluminescent B	1.1 mg/ml p-hydroxy coumaric acid in DMSO, light protected storage at RT
Coomassie	0.25 g coomassie Brilliant Blue G-250 in 100 ml ethanol, 100 ml $_{d}H_2O$, 25 ml acetic acid, light protected storage at RT
Coomassie destaining	100 ml ethanol, 100 ml $_{d}H_2O$, 25 ml acetic acid
Plasmid I	50 mM Tris-HCl (pH 7.5), 10 mM EDTA, 100 µg/ml RNaseA
Plasmid II	0.2 M NaOH, 1% (w/v) SDS
PMSF (100×)	17 mg/ml PMSF in isopropanol, light protected storage at −20°C
RNase A	1 mg/ml RNaseA in TE buffer, boiling 15 min in water bath, slowly cool down at RT, storage at −20°C
X-α-gal (1000×)	20 mg/ml in DMSO, light protected storage at −20°C

vectors containing the "Gateway attL/attR" sites. Zymoclean™ Gel DNA Recovery Kit (Zymo Research) was used to purify DNA fragments from agarose gels. The insertion of minor mutations within DNA fragments, was performed using the QuikChange® site-directed mutagenesis kit (Stratagene).

4.1.3 Buffers and solutions

All standard solutions and buffers used in this study were prepared as described in Ausubel *et al.* (1999) and Sambrook & Russell (2001) and are listed in Table 4.1 and Table 4.2. All additional specific solutions and buffers are listed in the corresponding method sections.

Table 4.2: Standard buffers used in this study.

Buffer (conc.)	Composition
Blotting (1×)	25 mM Tris base, 192 mM glycine, 0.3 % (w/v) SDS, pH 8.7, 20 % (v/v) methanol
Cracking (1×)	8 M urea, 5 % (w/v) SDS, 40 mM Tris-HCl (pH 5.6), 0.1 mM EDTA, 0.4 mg/ml bromphenol blue
gDNA extraction (1×)	0.1 M Tris-HCl (pH 7.5), 50 mM EDTA, 1 % (w/v) SDS, 0.5 M NaCl
GST (1×)	4.3 mM Na_2HPO_4, 1.47 mM KH_2PO_4, 0.137 M NaCl, 2.7 mM KCl, pH 7.3
GST elution (1×)	50 mM Tris-HCl, 0.046 g of reduced glutathione per 10 ml, pH 8.0
Laemmli (5×)	360 mM Tris-HCl (pH 6.8), 50 % (v/v) glycerol, 10% (w/v) SDS, 0.1 % bromphenol blue, 250 mM DTT
PBS (1×)	10 mM Na_2HPO_4, 1.8 mM KH_2PO_4, 0.14 M NaCl, 0.27 mM KCl, pH 7.3
PBST (1×)	PBS containing 0.05 %(v/v) of Tween20
PCR (1×)	10 mM Tris-HCl, 1.5 mM $MgCl_2$, 50 mM KCl, pH 8.3
TAE (10×)	48.8 g Tris base, 11.42 ml acetic acid, 7.44 g $Na_2EDTA \cdot 2\,H_2O$, pH 8.5
TE (1×)	10 mM Tris-HCl (pH 8.0), 1 mM EDTA
Tris-glycine (1×)	25 mM Tris base, 192 mM glycine, 0.1% (w/v) SDS, pH 8.7

4.1.4 Enzymes

All restriction enzymes, Taq DNA polymerase, T4 DNA ligase and Antarctic Phosphatase were obtained from New England Biolabs. Phusion® High-Fidelity DNA polymerase was obtained from Finnzymes and Pfu DNA polymerase from Thermo Scientific.

4.2 Oligonucleotides

All oligonucleotides used in this study were purchased from Eurofins MWG Operon GmbH and are listed in Table 4.3 and Table 4.4.

Non-target gene sequences such as restriction recognition sites and "CACC" for directional cloning of the PCR product in pENTR®, as well as nucleotide replacement sites within target gene sequences are underlined.

Table 4.3: Oligonucleotides used for cloning

Primer (ID #)	Sequence 5´- 3´	Tm (°C)
KPI106 overexpression under control of CaMV 35S promoter		
106-ORF-F (854)	CACCTAACAAATAATAAACCATGTCAATG	59.6
106-ORF-R (864)	AATTATTCTTTCTGGAAAACAAGAGG	56.9
KPI104 overexpression under control of CaMV 35S promoter		
104-ORF-F (887)	CACCTATGTCAATCAGATCCCTCAC	63.0
104-ORF-R (894)	GAATTCTCATTCTTTCTGGAACACAAC	60.4
RNAi target for CP knockdown		
CP-RNAi-F (1195)	CACCGGATTGAGCTTCCATGACT	62.4
CP-RNAi-R (1196)	CATAGGGTTGACCAGGCAATGT	62.4
RNAi target for SCP1 knockdown		
R-SCP-F (1518)	CACCATGAAGAAGGTTTCTCTTTATGCT	62.2
R-SCP-R (1519)	CATAGGGTTGACCAGGCAATGT	60.3
Removal of KPI106 stop codon		
106delS-F (1636)	GTTTTCCAGAAAGAAAAGGGTGGGCGCGC	69.5
106delS-R (1637)	GCGCGCCCACCCTTTTCTTTCTGGAAAAC	69.5
Removal of KPI104 stop codon		
104delS-F (1556)	GTGTTCCAGAAAGAAAAGGGTGGGCGCGC	70.9
104delS-R (1557)	GCGCGCCCACCCTTTTCTTTCTGGAACAC	70.9
CP eGFP fusion under control of CaMV 35S promoter		
CP-ORF-F (1517)	CACCATGGCACAGTGGACGCT	63.7
CP-w/o-R (1534)	GTTCGACGGTGAACCAGAGG	61.0
SCP1 eGFP fusion under control of CaMV 35S promoter		
SCP-ORF-F (1518)	CACCATGAAGAAGGTTTCTCTTTATGCT	62.2
SCP-w/o-R (1533)	GTTCGACGGTGAACCAGAGG	61.8
Amplification of free eGFP for localization control		
eGFP-F (1406)	CACCATGGTGAGCAAGGGCGA	63.7
eGFP-R (1405)	GTTACAAGTACAGCTCGTCCATG	60.6

Oligonucleotides for cloning continued

Primer (ID #)	Sequence 5´- 3´	Tm (°C)
KPI106 without signalpeptide for Y2H bait		
Nco1-106F (906)	CCATGGCTCAGTTTGTCTTGGAC	62.4
EcoR1-106R (890)	GAATTCTTATTCTTTCTGAAAACAAGAGG	61.3
KPI104 without signalpeptide for Y2H bait		
Nco1-104F (908)	CCATGGCTCAGTTTGTCATCGAC	62.4
EcoR1-104R (894)	GAATTCTCATTCTTTCTGGAACACAAC	60.4
KPI105 without signalpeptide for Y2H bait		
Nde1-105F (1638)	CATATGCAAATTGTCATAGACACAAGTGG	62.4
BamH1-105R (1639)	GGATCCTCATTCTTTCTGGACACAACT	63.7
KPI111 without signalpeptide for Y2H bait		
Nco1-111F (1640)	CCATGGCTCAGTTTGTCATCGACAC	64.6
111-3UTR (1658)	GTAAACTTCACCATCCTTCAGATTATTC	60.7
KPIc without signalpeptide for Y2H bait		
Nde1-KPIc-F (1648)	CATATGGCTAAACAAGTCTTAGACATACATG	62.9
BamH1-KPIc-R (1649)	GGATCCCTAAATAGACAAAGCACTATTTC	62.4
Substitution of K173G in KPI106		
106^{K173G}-F (1708)	AGAGGTATGTCCAAGTTGTGGGTTTGAAT GTGGAACTGTTG	72.4
106^{K173G}-R (1709)	CAACAGTTCCACATTCAAACCCACAACTT GGACATACCTCT	72.4
Substitution of FE174VQ in KPI106		
$106^{FE174VQ}$-F (1853)	GGTATGTCCAAGTTGTAAAGTACAATG TGGAACTGTTGATATG	69.4
$106^{FE174VQ}$-R (1854)	CATATCAACAGTTCCACATTGTACTTT ACAACTTGGACATACC	69.4
Substitution of K170G in KPI104		
104^{K170G}-F (1710)	CAGAGGCTTGTCCAAGTTGTGGAGTACA ATGTGGGACTGTTG	74.3
104^{K170G}-R (1711)	CAACAGTCCCACATTGTACTCCACAACT TGGACAAGCCTCTG	74.3
Substitution of VQ171FE in KPI104		
$104^{VQ171FE}$-F (1851)	CTTGTCCAAGTTGTAAGTTTGAA TGTGGGACTGTTGGTG	70.5
$104^{VQ171FE}$-R (1852)	CACCAACAGTCCCACATTCAAA CTTACAACTTGGACAAG	70.5
CP without signalpeptide for Y2H prey		
Nde1-CP-F (1209)	CATATGAGCTTCCATGACTCCAATC	61.3
Nde1-CP-R (1210)	CATATGTTAGGCAACAACAGG ATAGG	61.6

Oligonucleotides for cloning continued

Primer (ID #)	Sequence 5´- 3´	Tm (°C)
SCP1 without signalpeptide for Y2H prey and bait		
SCP1_w/oSP-f (1840)	AGTCAAGCTGATAAATTCAATGAG TTTATTCT	62.0
Xho1-SCP1-r (1457)	CGCTCGAGTTAATTCGACGGTGA ACC	66.4
SbtM1 without signalpeptide for Y2H prey		
Nde1-SbtM1-f (1204)	CATATGGGATTGTGGAGCGACGA TAATC	65.1
Nde1-SbtM1-r (1206)	CATATGGCTGCAGAGGAGAATCTA GAAC	65.1
SCP2 without signalpeptide and autoinhibitory domain for Y2H prey		
Sfi1-SCP2_for (1855)	GGCCATGGAGGCCCCTGGTCAAC CATATG	62.1
Sfi1-SCP2_rev (1856)	GGCCGAGGCGGCCTTAGTTTGATGG TGAAAC	58.1
Substitution of S226A in SCP1		
$SCP1^{S226A}F$ (1861)	GATTTTTACATAACTGGAGAGGCT TATGCCGGTCATTATGTTCC	71.3
$SCP1^{S226A}R$ (1862)	GGAACATAATGACCGGCATAAGC CTCTCCAGTTATGTAAAAATC	71.3
Substitution of D191A and Y193A in SCP1		
$TAD_{mut}F$ (1974)	GATAAGTCCACTGCTAAAGCTGCC GCTGTCTTCCTAATCAAC	73.3
$TAD_{mut}R$ (1975)	GTTGATTAGGAAGACAGCGGCAGC TTTAGCAGTGGACTTATC	73.3
Pulldown assay		
EcoR1-106-F (1280)	GAATTCGCTCAGTTTGTCTTG GAC	61.0
Sal1-106-R (1281)	GTCGACTTATTCTTTCTGGAAAACAA GAG	62.4
EcoR1-CP-F (1282)	GAATTCGGATTGAGCTTCCATGACTC	63.2
Sal1-CP-R (1283)	GTCGACTTAGGCAACAACAGGATAG	63.0

Table 4.4: Oligonucleotides used for qPCR

Primer name (ID #)	Sequence 5´- 3´	Tm (°C)	Amplicon (bp)
qPCR KPI106			
KPI106_F1	TCAAATTCACACCATTTGCTC	54	146
KPI106_R1	CCTTCTTCCACTCTTGGTGTC	59.8	
qPCR KPI104 (amplicon in 3´UTR)			
KPI104_F1 (1626)	CTTGAGAGTAACATTCCTAACCTC	56	392
KPI104_re (1589)	GCTACACAAGCTTATTAACGATG	56	
qPCR KPI104 (amplicon in ORF)			
KPI104_F1 (1626)	CTTGAGAGTAACATTCCTAACCTC	56	310
KPI104_R3	TCATTCTTTCTGGAACACAACTG	56	
qPCR CP			
CP-qPCR_F2	GAGGATGAATTGAAACATGCAG	56.5	260
CP-qPCR_R2	CACATATTCTTCCCCATTTCCA	56.5	
qPCR SCP			
SCP-qPCR-F1	CATTGTGTCATGACTCTTCTCT	56.5	273
SCP-qPCR-R1	GAACCACTACATCAGTATCACCA	58.9	
qPCR PT4			
MtPT4-F1	GTGCGTTCGGGATACAATACT	57.9	200
MtPT4-R1	GAGCCCTGTCATTTGGTGTT	57.3	
qPCR MtTEF			
MtTEF-F3	TACTCTTGGAGTGAAGCAGATG	56	250
MtTEF-R2	GTCAAGACCCTCAAGGAGAG	56	
qPCR RiTEF			
RiTEF-F	TGTTGCTTTCGTCCCAATATC	56	177
RiTEF-R	GGTTTATCGGTAGGTCGAG	56	
qPCR RiMST2			
RiMST2-F	GGCAGGATATTTGTCTGATAG	56	100
RiMST2-R	GCAATAACTCTTCCCGTATAC	56	
qPCR KPI105			
KPI105-F (1625)	CTTGAGAGTAACATTCCAAGGTTT	57.6	406
KPI105-R (1590)	GCTTCACATATATATGTTTATTAA ACTTAC	58.9	
qPCR KPI111			
KPI111-F (1844)	GAATAATCTGAAGGATGGTGAAG	57.4	93
KPI111-R (1843)	GATGACTCCATATTTATTGAGTAA ACA	57.1	

Table 4.4: Oligonucleotides used for qPCR

Primer name (ID #)	Sequence 5´- 3´	Tm (°C)	Amplicon (bp)
qPCR KPIc			
KPIc-F (1845)	GTGCTTTGTCTATTTAGTATGAGT	55.9	94
KPIc-F (1846)	CACAAAAGCAAGTGTAGTTATTA TTC	56.9	

4.3 Plasmids and plasmid constructs

4.3.1 Plasmids

- **pCR®II-TOPO®** (Invitrogen): Used for cloning of PCR products containing 3´A-overhangs. Vector contains an ampicillin (100 mg/l) and kanamycin (50 mg/l) resistance.

- **pENTR™/D-TOPO®** (Invitrogen): Used for directional cloning of blunt-end PCR products. Plasmid contains attL sites for subsequent Gateway® cloning of PCR products in various destination vectors containing attR sites. Vector contains a kanamycin (50 mg/l) resistance.

- **pGBKT7** (Clontech): Expression of Gal4 DNA-binding domain fusion protein (bait) with N-terminal c-myc-tag under control of the ADH1 promoter for yeast two-hybrid analyses. Provides kanamycin (50 mg/l) resistance in *E. coli* and tryptophan (TRP1) nutritional marker in yeast.

- **pGADT7-Rec** (Clontech): Expression of Gal4 DNA-activation domain fusion protein (prey) with N-terminal HA-tag under control of the ADH1 promoter for yeast two-hybrid analyses. Provides ampicillin (100 mg/l) resistance in *E. coli* and leucin (LEU2) nutritional marker in yeast.

- **pGBKT7_p53** (Clontech): Positive control plasmid, containing murine p53 fused to the Gal4 DNA-binding domain.
- **pGBKT7_Lam** (Clontech): Negative control plasmid that encodes a fusion of the human lamin C protein and the Gal4 DNA-binding domain.
- **pGADT7-Rec_T** (Clontech): Control plasmid, contains a fusion of the SV40 large T antigen and the Gal4 DNA-activating domain.
- **pET28a** (Novagen): Expression of N- and C-terminal 6×His-tag fusion proteins under control of the T7 promoter. Contains a kanamycin (50 mg/l) resistance.
- **pGEX4T1** (GE Healthcare): Expression of N-terminal (GST) fusion proteins under control of the tac promoter. Contains an ampicillin (100 mg/l) resistance.
- **pCGFP-RR** (Kamiri *et al.*, 2002): Binary Gateway destination vector for *Agrobacterium* mediated plant transformation that was used for expression of proteins under control of the CaMV 35S promoter. Proteins were expressed either containing their endogenous stop codon or as C-terminal eGFP fusion when the stop codon was removed. The vector provides spectinomycin (100 mg/l) resistance in bacteria and transformed roots are kanamycin resistant (25 mg/l). Furthermore, the vector contains a DsRed cassette under control the ubiquitin promoter for identification of transformed plant tissue (Kuhn *et al.*, 2010).
- **pK7GWIWG2D(II),0** (Kamiri *et al.*, 2002): Binary Gateway destination vector for *Agrobacterium* mediated plant transformation. Used for expression of hairpin RNA under control of the CaMV 35S

promoter for post-transcriptional gene silencing. Vector provides spectinomycin (100 mg/l) resistance in bacteria and transformed roots are kanamycin resistant (25 mg/l). The vector also contains a p35S:GFP cassette for identification of transformed plant tissue. In the following, plasmids with pK7GWIWG2D(II) backbone are referred to as pRNAi.

4.3.2 Plasmid constructs generated in this study

4.3.2.1 Overexpression and localization studies

- **pCGFP-RR_106:** *KPI106* ORF including stop codon was amplified from cDNA (primers #854 and #864), cloned into pENTR™/D-TOPO®, and subcloned into the binary Gateway vector pCGFP-RR. **pCGFP-RR_KPI104** was created accordingly, using primers #887 and #894.

- **pCGFP-RR_106GFP:** After deletion of the stop codon of *KPI106* ORF present in pENTR™/D-TOPO® using site-directed mutagenesis (primers #1636 and #1637), subcloning into the binary vector pCGFP-RR was carried out. **pCGFP-RR_104GFP** was created accordingly, using primers #1556 and #1557.

- **pCGFP-RR_SCP1GFP:** *SCP1* ORF without stop codon was amplified from cDNA (primers #1518 and #1533), cloned into pENTR™/D-TOPO® and subcloned into the binary gateway vector pCGFP-RR. **pCGFP-RR_CPGFP** (primers #1517 and #1534) and **pCGFP-RR_eGFP** (primers #1406 and #1405) were created accordingly.

4.3.2.2 RNAi-mediated silencing

- **pRNAi_SCP:** A 259 bp fragment (+1 to +260) of the *SCP1* ORF was amplified from cDNA (primers #1518 and #1519), cloned into pENTR™/D-TOPO®, and subcloned into pK7GWIWG2D(II),0 by LR-recombination.

- **pRNAi_CP:** A 324 bp fragment (+121 to +445) of the *CP* ORF was amplified (primers #1195 and #1196) and cloned accordingly.

4.3.2.3 Yeast two-hybrid analyses

- **pGBKT7_106:** *KPI106* ORF without secretion signal peptide was amplified from cDNA with primers containing restriction sites (#906 and #890), cloned into pCR®II-TOPO®, digested with the respective restriction enzymes and ligated into the bait vector pGBKT7. **pGBKT7_104** (primers #908 and #894)**, pGBKT7_105** (primers #1638 and #1639), **pGBKT7_111** (primers #1640 and #1658) and, **pGBKT7_KPIc** (primers #1648 and #1649) were created accordingly.

- **pGADT7-Rec_CP:** *CP* ORF without secretion signal peptide was amplified from cDNA with primers containing restriction sites (#1209 and #1210), cloned in pCR®II-TOPO®, digested with the respective restriction enzymes and ligated into the prey vector pGADT7-Rec. **pGADT7-Rec_SbtM1** (primers #1204 and #1206) and **pGADT7-Rec_SCP2** (primers #1855 and #1856) were created accordingly. For **pGADT7-Rec_SCP1**, *SCP1* ORF was amplified without secretion signal peptide using primers primers #1840 and #1457, cloned into pCR®II-TOPO®, cut out using EcoR1 and Xho1, and ligated into pGADT7-Rec.

- **pGBKT7_106$^{\mathbf{K173G}}$:** Derivative of pGBKT7_106 carrying the amino acid substitution K173G. For site-directed mutagenesis, the primers #1708 and #1709 were used. **pGBKT7_104$^{\mathbf{K170G}}$** (primers #1710 and #1711), **pGBKT7_106$^{\mathbf{FE174VQ}}$**(primers #1853 and #1854), and **pGBKT7_104$^{\mathbf{VQ171FE}}$**(primers #1851 and #1852) were created accordingly.

- **pGBKT7_SCP1$^{\mathbf{S226A}}$:** SCP1 was cut out of pCR®II-TOPO® (see also pGADT7-Rec_SCP1) using EcoR1 and Xho1, and ligated into pGBKT7. Substitution of S226A was carried out by site-directed mutagenesis with primers #1861 and #1862.

- **pGADT7-Rec_SCP1$^{\mathbf{S226A}}$:** Substitution of S226A was carried out by site-directed mutagenesis with primers #1861 and #1862 using pGADT7-Rec_SCP1 as template.

- **pGBKT7_SCP1$^{\mathbf{S226A}}$TAD$_{\mathbf{mut}}$:** Substitutions of D191A and Y193A were carried out by site-directed mutagenesis with primers #1974 and #1975 using pGBKT7_SCP1^{S226A} as template.

4.3.2.4 Pulldown assays

- **pET28a_106:** *KPI106* ORF without secretion signal peptide and including stop codon was PCR amplified with primers containing restriction sites (#1280 and #1281) and cloned *via* restriction sites into pET28a. For **pET28a_104**, KPI104 was cut out of pCR®II-TOPO® (see also pGBKT7_104) with BamH1 and Xho1, and ligated into pET28a.

- **pGEX4T1_CP:** *CP* ORF without secretion signal peptide was PCR amplified with primers containing restriction sites (#1282 and #1283) and cloned *via* restriction sites into pGEX4T1.

4.4 Organisms

4.4.1 *Rhizophagus irregularis*

For mycorrhization assays, the AM fungus *R. irregularis*, corresponding to *Glomus intraradices* DAOM 181602 (Schenck & Smith, 1982), was used.

4.4.2 *Saccharomyces cerevisiae*

The *S. cerevisiae* strain AH109 was obtained from the Matchmaker™ Library Construction & Screening Kit (CLONTECH) and contains the genotype {MATa, trp1-901, leu2-3, 112, ura3-52, his3-200, gal4Δ, gal80Δ, LYS2::GAL1$_{UAS}$-GAL1$_{TATA}$-HIS3, GAL2$_{UAS}$-GAL2$_{TATA}$-ADE2, URA3::MEL1$_{UAS}$-MEL1$_{TATA}$-lacZ}.

4.4.3 *Escherichia coli*

For general cloning work the *E. coli* strains **XL1 Blue** {recA1 endA1 gyrA96 thi-1 hsdR17 supE44 relA1 lac[F´pro AB laclqZΔM15 Tn10(tetr)]} (STRATAGENE) and **Mach1** {F– Φ80lacZΔM15 ΔlacX74 hsdR(rK–, mK+) ΔrecA1398 endA1 tonA} (INVITROGEN) were used. For recombinant protein expression, the *E.coli* **BL21**(DE) pLysS strain {F- ompT gal dcm lon hsdSB(rB- mB-) λ(DE3) pLysS(cmR)} was used.

4.4.4 *Agrobacterium* strains

For transgenic *M. truncatula* hairy root production, the *A. rhizogenes* strain *ARquaI* (Quandt *et al.*, 1993) and for transient transformation of *Nicotiana benthamiana* leaves, the *A. tumefaciens* strain GV3101 Koncz & Schell (1986) were used. Table 4.5 contains a detailed list including genotypes and references of all *Agrobacterium* strains used in this work.

Table 4.5: *Agrobacterium* strains used in this work

Strain	Genotype	Reference
A. rhizogenes:		
ARquaI	Ri-plasmid: pRiA4b (wt), Sm^r: 600 mg/l	Quandt *et al.* (1993)
ARquaI: pCGFP-RR_106	*ARquaI* transformed with the binary vector pCGFP-RR_106, Sm^r, Sp^r	This work
ARquaI: pCGFP-RR_104	*ARquaI* transformed with the binary vector pCGFP-RR_104, Sm^r, Sp^r	This work
ARquaI: pCGFP-RR	*ARquaI* transformed with the binary vector pCGFP-RR, Sm^r, Sp^r	This work
ARquaI: pRNAi_CP	*ARquaI* transformed with the binary vector pRNAi_CP, Sm^r, Sp^r	This work
ARquaI: pRNAi_SCP	*ARquaI* transformed with the binary vector pRNAi_SCP, Sm^r, Sp^r	This work
ARquaI: pRNAi	*ARquaI* transformed with the binary vector pRNAi, Sm^r, Sp^r	This work
ARquaI: pCGFP-RR_106GFP	*ARquaI* transformed with the binary vector pCGFP-RR_106GFP, Sm^r, Sp^r	This work
ARquaI: pCGFP-RR_104GFP	*ARquaI* transformed with the binary vector pCGFP-RR_104GFP, Sm^r, Sp^r	This work
ARquaI: pCGFP-RR_SCP1GFP	*ARquaI* transformed with the binary vector pCGFP-RR_SCP1GFP, Sm^r, Sp^r	This work
A. tumefaciens:		
GV3101	Ti-plasmid pMP90 (pTiC58 ΔT-DNA), Gm^r: 40 mg/l	Koncz & Schell (1986)
GV3101: pCGFP-RR_106GFP	GV3101 transformed with the binary vector pCGFP-RR_106GFP, Gm^r, Sp^r	This work
GV3101: pCGFP-RR_104GFP	GV3101 transformed with the binary vector pCGFP-RR_104GFP, Gm^r, Sp^r	This work
GV3101: pCGFP-RR_CPGFP	GV3101 transformed with the binary vector pCGFP-RR_CPGFP, Gm^r, Sp^r	This work
GV3101: pCGFP-RR_SCP1GFP	GV3101 transformed with the binary vector pCGFP-RR_SCP1GFP, Gm^r, Sp^r	This work
GV3101: pCGFP-RR_eGFP	GV3101 transformed with the binary vector pCGFP-RR_eGFP, Gm^r, Sp^r	This work
GV3101:p19	GV3101 transformed with the binary vector p19, Gm^r, Km^r	K. Harter (Tübingen)

4.4.5 *Daucus carota*

A. rhizogenes transformed *D. carota* hairy roots were used for cultivation of *R. irregularis* in monoxenic culture (Table 4.6).

4.4.6 *Medicago truncatula*

All experiments were carried out using the *M. truncatula* L. Gaertn. var. Jemalong A17 cultivar (Perkiss Seeds, Australia). Transgenic *M. truncatula* hairy root lines were created as described in Section 5.2.1 and are listed in Table 4.6.

Table 4.6: Transgenic hairy roots used in this work

Root line	Genotype	Reference
D. carota:		
DcARqualI	transformed with *ARqualI*	No data
M. truncatula:		
MtARqualI	transformed with *ARqualI*	Kuhn *et al.* (2010)
p35S:106-1, p35S:106-2	transformed with *ARqualI*:pCGFP-RR_106, DsRed, Km^r	This work
p35S:104-1, p35S:104-2	transformed with *ARqualI*:pCGFP-RR_104, DsRed, Km^r	This work
p35S_EVctrl	transformed with *ARqualI*:pCGFP-RR, DsRed, Km^r	This work
R-CP_1, R-CP_2	transformed with *ARqualI*:pRNAi_CP, GFP, Km^r	This work
R-SCP_1, R-SCP_2	transformed with *ARqualI*:pRNAi_SCP, GFP, Km^r	This work
R_EVctrl	transformed with *ARqualI*:pRNAi, GFP, Km^r	This work
p35S:106GFP	transformed with *ARqualI*:pCGFP-RR_106GFP, Km^r	This work
p35S:104GFP	transformed with *ARqualI*:pCGFP-RR_104GFP, Km^r	This work
p35S:SCP1GFP	transformed with *ARqualI*:pCGFP-RR_SCP1GFP, Km^r	This work

4.4.7 *Nicotiana benthamiana*

As a heterologous system for localization of *M. truncatula* proteins, the tobacco species *N. benthamiana* was used.

4.5 Culture media

4.5.1 Media for cultivation of *M. truncatula* and *R. irregularis*

4.5.1.1 Minimal (M) Medium (Bécard & Fortin, 1988)

Table 4.7: Ingredients of M medium

Ingredient	Stock solution	Final concentration	per l medium
Macroelements			
KNO_3	3.2 g/l	80 mg/l	
$MgSO_4 \cdot 7\,H_2O$	29.24 g/l	731 mg/l	25 ml
KCl	2.6 g/l	65 mg/l	
KH_2PO_4	48 mg/l	4.8 mg/l	100 ml
$Ca(NO_3)_2$	115.2 mg/l	288 mg/l	2.5 ml
NaFe-EDTA	3.2 g/l	8 mg/l	2.5 ml
Microelements			
$MnCl_2 \cdot 4\,H_2O$	6 g/l	6 mg/l	
H_3BO_3	1.5 g/l	1.5 mg/l	1 ml
$ZnSO_4 \cdot 7\,H_2O$	2.65 g/l	2.65 mg/l	
Other microelements			
$NaMoO_4 \cdot 2\,H_2O$	24 mg/l	2.4 µg/l	100 µl
$CuSO_4 \cdot 5\,H_2O$	1.3 g/l	0.3 mg/l	
Vitamines			
Glycine	0.3 g/l	3 mg/l	
Myo-inositol	5 g/l	50 mg/l	
Nicotic acid	50 mg/l	0.5 mg/l	10 ml
Pyridoxine HCl	10 mg/l	0.1 mg/l	
Thiamine HCl	10 mg/l	0.1 mg/l	

The pH was adjusted to 5.5. For *in vitro* culture of roots 10 g/l D-Saccharose was added (medium is then referred to as MS). Solid medium contained 0.3 % (w/v) phytagel (SIGMA).

4.5.1.2 Modified Fåhræus Medium

Table 4.8: Ingredients of modified Fåhræus medium

Ingredient	Stock solution	Final concentration	per l medium
Macronutrients			
$CaCl_2$	1 M	1 mM	1 ml
$MgSO_4$	0.5 M	0.5 mM	1 ml
KH_2PO_4	0.7 M	0.7 mM	1 ml
NaH_2PO_4	0.4 M	0.4 mM	2 ml
Fe-Citrate	20 mM	20 µM	1 ml
NH_4NO_2	1 M	0.5 mM	0.5 ml
Micronutrients			
$MnCl_2$	1 mg/ml	100 µg/ml	100 µl
$CuSO_4$	1 mg/ml	100 µg/ml	100 µl
$ZnCl_2$	1 mg/ml	100 µg/ml	100 µl
H_3BO_3	1 mg/ml	100 µg/ml	100 µl
$NaMoO_4$	1 mg/ml	100 µg/ml	100 µl

The pH was adjusted to 6.5 and solid medium contained 1.5 % (w/v) agar.

4.5.2 Bacteria and yeast culture media

All media were prepared with dH_2O and autoclaved. Solid media contained 1.5 % (w/v) agar. If required, antibiotics, glucose, and other reagents were added after autoclaving and cooling down to 55°C.

- **LB medium:** 10 g/l tryptone, 5 g/l yeast extract, 10 g/l NaCl, pH 7.0
- **SOB medium:** 20 g/l tryptone, 5 g/l yeast extract, 0.584 g/l NaCl, 0.18 g/l KCl, pH 7.0, autoclave, add 10 ml/l of sterile Mg^{2+} solution (1 M $MgCl_2 \cdot 6\,H_2O$, 1 M $MgSO_4 \cdot 7\,H_2O$)
- **SOC medium:** SOB medium supplied with 20 mM glucose
- **YPDA medium:** 20 g/l peptone, 10 g/l yeast extract, 15 ml of 0.2 % (w/v) adenine hemisulfate, 0.1 M glucose, 2 % (w/v) agar, pH 6.5

- **SD medium:** 6.7 g/l yeast nitrogen base without amino acids (FORMEDIUM™), 100 ml/l of 10× LWHA dropout solution, 10 ml/l of 100× stock solution of each required dropout supplement, 0.1 M glucose, pH 5.8, 2 % (w/v) agar

- **H_2O-agar:** 0.8 % (w/v) agar in dH_2O

4.5.3 Supplement solutions for bacteria and yeast culture media

4.5.3.1 Antibiotic solutions

Antibiotics were prepared as 1000× stock solutions (Table 4.9), dissolved in dH_2O or $ethanol_{abs}$, sterile filtered (0.22 µm), and stored at −20°C.

Table 4.9: Antibiotic solutions

Antibiotic	1000× stock [mg/ml]
Ampicillin (Amp)	100
Amoxclav®(HEXAL) (Aug)	400
Chloramphenicol (Cm)	40
Gentamycin (Gm)	40
Kanamycin (Km)	50
Spectinomycin (Sp)	100
Streptomycin (Sm)	600
Tetracyclin (Tet)	25

4.5.3.2 Dropout supplements for SD media

The nutrients for 10×LWHA dropout solution are listed in Table 4.10 and were dissolved in dH_2O, autoclaved, and stored at 4°C. Single dropout supplements were prepared as 100× stock solutions dissolved in dH_2O, autoclaved, and stored at 4°C (Table 4.11).

Table 4.10: Ingredients of 10×LWHA dropout solution

Nutrient	**Concentration [mg/l]**
L-Arginine HCl	200
L-Isoleucine	300
L-Lysine HCl	300
L-Methionin	200
L-Phenylalanine	500
L-Threonine	2000
L-Tyrosine	300
L-Uracil	200
L-Valine	1500

Table 4.11: Single dropout supplements

Nutrient	**100× stock [mg/ml]**
L-histidine	2
L-leucine	10
L-tryptophane	2
L-adenine hemisulfate	2

CHAPTER 5

METHODOLOGY

5.1 Microbiology methods and transformation protocols

5.1.1 Competent cell preparation and transformation of *E. coli*

500 ml of SOB medium containing the appropriate antibiotic were inoculated with a freshly prepared overnight culture to an OD_{600} of 0.2. Cells were grown at 37°C, 180 rpm to a density of OD_{600} of 0.6. The culture was transferred to sterile centrifuge tubes, incubated on ice for 10 min, and centrifuged at 4°C, 2,500 rpm, for 20 min. The supernatant was discarded and cells were resuspended in 100 ml of cold TB buffer (10 mM PIPES, adjust pH 6.7, add 55 mM $MnCl_2 \cdot 4H_2O$, 15 mM $CaCl_2 \cdot 2H_2O$, 250 mM KCl, sterilize with 0.22 µm filter) and incubated on ice for 10 min. The suspen-

sion was centrifuged at 4°C, 2,500 rpm for 20 min and the supernatant discarded. *E. coli* cells were resuspended in 10 ml of TB-DMSO (TB buffer containing 7 % (v/v) DMSO) and aliquots of 100 µl each in sterile 1.5 ml micro-centrifuge tubes were frozen in liquid nitrogen and stored at −80°C.

For transformation, one aliquot of chemically competent *E.coli* cells was thawed on ice for 10 min. Afterwards, plasmid DNA was added, gently mixed and incubated on ice for 30 min. *E.coli* cells were then heat shocked at 42°C for 1 min, immediately cooled on ice and 500 µl of SOC medium were added for recovery of the cells. The suspension was incubated at 37°C and 180 rpm for 60 min. Cells were centrifuged for 1 min at 13,000 rpm, resuspended in 100 µl of supernatant, plated on LB-agar containing the appropriate antibiotic(s) and incubated at 37°C overnight.

5.1.2 *E. coli* cultivation

E. coli was grown in LB liquid medium (Section 4.5.2) and incubated at 37°C, 180 rpm. On solid LB medium, *E. coli* was incubated under aerobic conditions at 37°C. For making frozen stocks, 750 µl of an exponentially growing culture were mixed with 375 µl of 20 % (v/v) glycerol and stored at −80°C.

5.1.3 Competent cell preparation and transformation of agrobacteria

500 ml of LB medium containing the appropriate antibiotic(s) were inoculated with a freshly prepared overnight culture to an OD_{600} of 0.2. Cells were grown at 28°C at 200 rpm to a density of OD_{600} of 0.6. The culture was transferred into sterile centrifuge bottles, incubated on ice for 15 min and centrifuged at 4°C, 2,500 rpm for 10 min. The supernatant was discarded, the pellets were resuspended in 250 ml of cold $_{dd}H_2O$ and

the suspension centrifuged at 4°C, 2,500 rpm for 10 min. This step was repeated two more times. The pellets were resuspended in 25 ml of cold 10 % (v/v) glycerol and centrifuged at 4°C, 2,500 rpm for 10 min. Afterwards, each pellet was resuspended in 2 ml of cold 10 % (v/v) glycerol and aliquots of 100 µl each in sterile 1.5 ml centrifuge tubes were frozen in liquid nitrogen and stored at −80°C.

For transformation, one aliquot of electro-competent cells was thawed on ice for 10 min. Afterwards, desalted plasmid DNA ("drop dialysis" technique; MILLIPORE "V" series membranes, 0.025 µm) was added, gently mixed, transferred into a 2-mm electroporation bulb (PEQLAB), and put on ice. Electroporation occurred at 2 V and 25 µF, and 500 µl of SOC medium were added immediately. Cell suspension was incubated at 28°C, 200 rpm for 2 hours. Afterwards, 100 µl of the suspension were plated on LB-agar containing the appropriate antibiotic(s).

5.1.4 Cultivation of agrobacteria

Liquid cultures of *Agrobacterium* strains were grown in LB medium (Section 4.5.2) containing the appropriate antibiotic(s) at 28°C and 220 rpm. On solid medium, *Agrobacterium* strains were incubated under aerobic conditions at 28°C. For making frozen stocks, 750 µl of an exponentially growing culture containing the appropriate antibiotic(s) were mixed with 375 µl of 20 % (v/v) glycerol and stored at −80°C.

5.1.5 Transformation of *S. cerevisiae*

The following method describes a modified version of the LiAc-mediated transformation protocol[1]. Therefore, 300 ml of YPDA medium were inoculated with a freshly prepared overnight culture to an OD_{600} of 0.2.

[1] Yeast protocols handbook (CLONTECH), PR742227, 2008

Cells were grown at 30°C at 220 rpm to a density of OD_{600} of 0.4–0.6. The culture was transferred to sterile centrifuge tubes and centrifuged at RT, 2,000 rpm for 5 min. The supernatant was discarded, cells were resuspended in 50 ml of sterile 1×TE buffer (pH 7.5), and centrifuged at RT, 2,000 rpm for 5 min.

For transformation, cells were resuspended in 1.5 ml of freshly prepared sterile TE/LiAc solution (1 ml of sterile 10×TE, 1 ml of sterile 1 M LiAc (pH 7.5 with acetic acid), 8 ml of sterile dH_2O). 0.1 µg of each plasmid, 0.1 mg of salmon sperm carrier DNA (Sigma), 100 µl of competent yeast cells and 600 µl of sterile PEG/LiAc solution (8 ml of 50 % (w/v) PEG 3350 in sterile dH_2O, 1 ml 10×TE, 1 ml of 1 M LiAc) were added to a sterile 1.5 ml centrifuge tube and mixed by vortexing at high speed for 10 sec. The suspension was incubated at 30°C, 220 rpm for 30 min. After adding 70 µl of DMSO and mixing by gentle inversion, cells were heat shocked at 42°C for 15 min. Tubes were chilled on ice for 1 min and centrifuged for 30 sec, 13,000 rpm at RT. The supernatant was discarded and cells were resuspended in 150 µl of sterile TE (pH 7.5) and plated on SD agar plates in absence of the respective auxotrophic nutrients.

5.1.6 *S. cerevisiae* cultivation

Liquid cultures of *S. cerevisiae* were grown at 30°C while shaking at 220 rpm. *S. cerevisiae* plates were incubated under aerobic conditions at 30°C. Untransformed AH109 strain was cultivated in YPDA medium (Section 4.5.2) and transformed strains were cultivated in SD medium lacking the appropriate auxotrophic nutrients for selection.

5.2 Working with plants and mycorrhization assay

5.2.1 Hairy root production

To obtain seedlings, *M. truncatula* seeds were incubated in sulfuric acid for 20 min until black spots appeared. After five washing steps with sterile $_dH_2O$, seeds were put on H_2O-agar (Section 4.5.2) and stored at 4°C overnight to synchronize for germination. The next day, seeds were incubated at 27°C in darkness to induce germination.

Hairy root production was performed by *Agrobacterium*-mediated transformation of *M. truncatula* germlings according to Boisson-Dernier *et al.* (2001). Therefore, the radicle of *M. truncatula* germlings was cut with a scalpel approximately 3 mm underneath the top and scratched over a bacterial lawn of *A. rhizogenes* strain *ARquaI* containing the desired binary vector. The infected germlings were put on modified Fåhræus (Table 4.8) slant plates, supplemented with 25 mg/l kanamycin to select for positive transformed hairy roots. Plates were sealed two times with PARAFILM®M and the root area was covered with aluminum foil. Plates were incubated in a plant growth chamber under constant 8 h darkness and 16 h light conditions at 25°C for 4 weeks until the transformed roots appeared. Transformed roots were excised and transferred on MS plates (Table 4.7) supplemented with 25 mg/l kanamycin (selection of transformed roots) and 400 mg/l Amoxclav® (elimination of *A. rhizogenes*) and incubated at 27°C in darkness for 4 weeks. Afterwards, roots were cultivated as described in Section 5.2.2.

5.2.2 *M. truncatula* hairy root cultivation

M. truncatula hairy roots were grown under sterile conditions on MS plates (Table 4.7) supplemented with 25 mg/l kanamycin if required, and incubated at 27°C in darkness (Bécard & Fortin, 1988). The petri dishes were sealed two times with PARAFILM®M (PECHINEY PLASTIC PACKAGING) to exclude contaminations and to prevent drying up. For continuous cultivation, root parts were transferred to fresh MS medium.

5.2.3 *R. irregularis* cultivation and the two-compartment plate system

R. irregularis was co-cultivated with *DcARqualI* hairy roots on MS plates (Table 4.7) at 27°C in darkness according to Bécard & Fortin (1988). Mycorrhized *D. carota* roots were regularly transferred to fresh MS plates under sterile conditions for steady cultivation.

The two-compartment plate system used in this study is based on St-Arnaud *et al.* (1996) and modified according to Kuhn *et al.* (2010). It consists of a single divided petri dish, whereas one compartment (proximal) contains M medium supplied with saccharose (MS), and the other one is filled with M medium lacking saccharose (referred to as distal compartment; Figure 5.1). Mycorrhizal *D. carota* roots were placed in the proximal compartment close to the plate divider, plates were sealed with PARAFILM®M and incubated at 27°C in darkness for 8 weeks. In this way, growth of mycorrhizal *D. carota* hairy roots is suppressed in the distal compartment due to the lack of a carbon source, whereas fungal hyphae can cross the divider and sporulate within the distal side. When spores were visible on the distal side (8 weeks), M medium was removed and subsequently refilled with MS containing 2.4 mg/l of KH_2PO_4 (half the concentration of phosphate source compared to the conventional MS medium ($\frac{1}{2}PO_4{}^{3-}$);

Figure 5.1: Two-compartment plate set-up for mycorrhization assays. The proximal compartment contains the monoxenic culture of *R. irregularis* and *D. carota* hairy roots, whereas the distal side is filled with low-phosphate medium forcing the *M. truncatula* hairy roots into AM symbiosis with *R. irregularis.* Fungal hyphae cross-over into the distal compartment is facilitated by a poured step (red dashed line), whereas the cellophane layer (blue dashed line) compacts the growth of fungal hyphe close to the *M. truncatula* hairy roots.

the low concentration of phosphate forces the *M. truncatula* hairy roots to engage in AM symbiosis). Next, commercially available cellophane foil (pre-treatment: cut in shape of a half compartment, put in $_dH_2O$ and autoclaved two times) was placed on top of the fresh MS medium in the distal compartment, and covered with another layer of MS ($\frac{1}{2}PO_4{}^{3-}$). Then, plates were put angular in order to pour a step decreasing from the plate divider towards the distal compartment. Plates were sealed with PARAFILM®M and incubated at 27°C in darkness for ten days. Meanwhile, the fungal hyphae can colonize the freshly prepared distal compartment, whereas the cellophane foil leads a compact growth of the hyphal network. After ten days, the plates are ready to use for mycorrhization assays.

For mycorrhization assays, the respective *M. truncatula* hairy roots were put on top of the MS ($\frac{1}{2}PO_4{}^{3-}$) medium within the distal compartment, plates were sealed with PARAFILM®M, and incubated in darkness at 27°C for the desired time of mycorrhization.

5.2.4 Transient transformation of *Nicotiana benthamiana* leaves

Five week old tobacco plants were infiltrated with *A. tumefaciens* strains containing the desired binary expression vector, according to Schütze *et al.* (2009). Therefore, a 25 ml flask containing 5 ml LB medium and antibiotics was inoculated with 1 ml of a well grown *A. tumefaciens* overnight culture and incubated at 28°C, 200 rpm for 3 hours. The cultures were put in 15 ml centrifuge tubes and centrifuged at RT for 5000 rpm for 15 min. The supernatant was discarded and pellets were resuspended in 1 ml of AS medium (10 mM MES-KOH (pH 5.6), 10 mM $MgCl_2$, 200 µM acetosyringon). Then, the OD of each *A. tumefaciens* culture was adjusted to OD_{600} of 0.7–0.8 with AS medium. Each *A. tumefaciens* culture was mixed with an *A. tumefaciens* culture containing a binary vector for expression of the p19 silencing inhibitor (Voinnet *et al.*, 2003) in a 1:1 ratio and incubated on ice for 2 hours. Afterwards, *A. tumefaciens* cultures were infiltrated into the abaxial air space of the tobacco leaves using a syringe without a needle. Three days post-infection, the infiltrated leaves were used for microscopy or protein extraction.

5.3 Staining, microscopy and quantification methods

5.3.1 Ink and vinegar staining

This method is based on a protocol of Vierheilig *et al.* (1998). Mycorrhizal *M. truncatula* hairy roots were treated with 10 % (w/v) KOH at 80°C for 45 min to clear the tissue. Then, roots were washed three times for each 3 min with tap water at RT. Roots were stained with 5 % (v/v) ink (Pelikan) in commercial vinegar (containing 5 % acid) for 3 min at RT, and washed three times for 3 min with destaining solution (10 % (v/v) vinegar in $_{d}H_2O$). Roots were incubated in destaining solution overnight at RT with agitation. Then, destained roots were mounted on a microscope slide with glycerol, sealed with clear nail polish and stored at RT.

5.3.2 WGA-fluorescein staining

For phenotypical analyses, mycorrhizal hairy roots were treated with 10 % (w/v) KOH for 45 min and neutralized with 2 % (v/v) HCl. After three washes with PBS, roots were put into staining solution (PBS, 10 µg/ml Wheat Germ Agglutinin (WGA)-Fluorescein, 0.02 % (v/v) Tween20) overnight at 4°C with gentle agitation. Samples were stored in PBS at 4°C. For microscopy, stained roots were mounted on a microscope slide in PBS and analyzed within 30 min after preparation.

5.3.3 Microscopy

Ink stained mycorrhizal hairy roots were analyzed using an Axio Imager Z1 microscope (Zeiss) in the brightfield channel with AxioVision 4.7 software (Zeiss).

WGA-fluorescein stained mycorrhizal hairy roots were analyzed using a Leica TCS SP5 confocal microscope with excitation at 488 nm and emission collected from 505 nm to 520 nm.

Leica TCS SP5 was also used to visualize heterologous eGFP fusion proteins expressed in *N. benthamiana* leaves. Therefore, the epidermal cell layer of the lower leaf surface was peeled off and transferred on a microscope slide in PBS. Visualization occurred immediately after preparation with excitation at 488 nm and emission collected from 505 nm to 520 nm. Free DsRed was excited at 488 nm and emission collected from 610 nm to 650 nm. All microscopic images in this thesis represent maximal projections of z-stack images.

5.3.4 Quantification of mycorrhizal colonization

Quantification of AM colonization of ink stained mycorrhizal hairy roots was carried out according to Trouvelot *et al.* (1986) using the program MYCOCALC[2]. Between 60 and 100 fragments of each mycorrhizal hairy root line were analyzed and figures in this thesis are representative of average results.

5.4 Molecular biology methods

Standard molecular biology methods, such as purification of DNA, addition of 3´A-overhangs or molecular cloning techniques, are followed protocols described in Ausubel *et al.* (1999) and Sambrook & Russell (2001). The integrity of nucleic acids was analyzed by electrophoresis in 0.5×TAE with a Mupid-ex U electrophoresis system (EUROGENTEC). Concentration of nucleic acids was quantified by photometry using a NanoDrop® ND-1000 spectrophotometer, whereas purity was determined by the ratio of absorbance at 260 nm and 280 nm (A_{260}/A_{280}). For purified DNA samples, the ratio was ~1.8, for purified RNA samples A_{260}/A_{280} was ~2.0.

5.4.1 *E. coli* plasmid preparation

E. coli plasmid preparation was carried out using the alkaline lysis method according to Sambrook & Russell (2001). Therefore, *E. coli* cultures were transferred into 2 ml centrifuge tubes and centrifuged for 1 min at 13,000 rpm. The supernatant was discarded and cell pellets resuspended in 200 µl of solution I (Table 4.1). For cell lysis, 200 µl of solution II (Table 4.1) were added and mixed by gentle inversion. To stop the lysis reaction, 200 µl of 1.5 M KAc (pH 4.8) were added, gently mixed by inversion and

[2]www2.dijon.inra.fr/mychintec/Mycocalc-prg/download.html

centrifugation for 15 min at 13,000 rpm followed. The supernatant was transferred to a 1.5 ml centrifuge tube containing 600 µl isopropanol to precipitate plasmid DNA. After 2 min of incubation at RT, samples were centrifuged for 15 min at 13,000 rpm. The supernatant was discarded, and the DNA pellet was washed with 500 µl of 70 % (v/v) ethanol, and dried on a heating block at 65°C for 10 min. The DNA pellet was dissolved in 50 µl of TE buffer at 65°C for 15 min and samples were stored at −20°C. Plasmid DNA was analyzed by restriction digest and sequencing.

5.4.2 Isolation of genomic DNA from *M. truncatula* hairy roots

For each gDNA sample, hairy roots of three fully grown MS plates (Table 4.7) were blotted dry with a paper towel and ground with pestle and mortar under liquid nitrogen. Subsequently, the root powder was transferred into a 50 ml centrifuge tube (SARSTEDT) containing 6 ml of extraction buffer (Table 4.2) and incubated in a water bath at 65°C for 30 min. Then, 2.5 ml of KAc solution (for 100 ml: 60 ml of 5 M KAc, 11.5 ml of acetic acid, 28.5 ml of H_2O, storage RT) were added and incubated on ice for 30 min. Next, the samples were centrifuged at 5,000 rpm, 4°C for 30 min and the supernatant was transferred into a new 50 ml centrifuge tube. Then, an equal volume of isopropanol was added and samples were incubated for 10 min at RT and centrifuged at 5,000 rpm, 4°C for 20 min. The supernatant was discarded and the pellet dried at 65°C. The pellet was dissolved in 100 µl of TE buffer, and 5 µl of RNase A solution (Table 4.1) were added. Samples were incubated for 1 hour at 37°C and stored at 4°C. To check the integrity of genomic DNA, 2 µl of each sample were loaded on an 0.8 % agarose gel.

5.4.3 Isolation of total RNA from *M. truncatula* hairy roots

All solutions used for work with RNA were prepared with Milli-Q™(MILLIPORE) water pre-treated with DEPC (H_2O_{DEPC}: 0.1 % (v/v) DEPC in Milli-Q™ water, stirred over night and autoclaved).

Isolation of total RNA of hairy roots was carried out using the TRIzol® (INVITROGEN) method as described by the supplier. Therefore, roots were homogenized in 1.5 ml centrifuge tubes under liquid nitrogen using plastic pestles (Rotilabo®micro-homogenizers, ROTH). The root powder was resuspended in 1 ml of TRIzol® and incubated at RT for 5 min. Then, 200 µl of Chloroform were added, samples were shaken by hand for 15 s, incubated at RT for 3 min and centrifuged at 4°C, for 15 min at 12,000 rpm. The mixture has then separated into three phases, whereas 500 µl of the upper aqueous phase (contains the RNA) were transferred into a new 1.5 ml centrifuge tube. Next, 500 µl of isopropanol were added and samples were incubated for 10 min at RT. Centrifugation at 4°C for 10 min at 12,000 rpm followed and the supernatant was discarded. The pellet was washed with 1 ml of 75% (v/v) ethanol and centrifuged at 4°C for 5 min at 8,000 rpm. The supernatant was discarded, the RNA pellet air-dried and dissolved in 90 µl of H_2O_{DEPC} at 65°C for 10 min. The samples were centrifuged at 4°C for 1 min at 12,000 rpm to remove debris.

In the following, a modified protocol from Untergasser (2008)[3]) was used to precipitate the RNA with LiCl: Therefore, the supernatant obtained in the last centrifugation step was transferred into a fresh 1.5 ml centrifuge tube, 30 µl of a 8 M LiCl solution were added and mixed well. The RNA was precipitated by incubation at −20°C for 1 hour, and samples were centrifuged at 4°C for 30 min at 12,000 rpm. The supernatant was discarded and the pellet washed with 500 µl of 75 % (v/v) ethanol and centrifuged at

[3] www.untergasser.de/lab/protocols/miniprep_rna_ctab_v1_0.htm

4°C for 2 min at 12,000 rpm. The RNA pellet was air-dried and dissolved in 25 µl of H_2O_{DEPC} at 65°C for 10 min. Then, samples were chilled on ice and 0.3 µl of RNaseOUT™ were added. Integrity of the RNA was confirmed on an 1.5 % (w/v) agarose gel. Samples were stored at −80°C.

5.4.4 Polymerase Chain Reaction (PCR)

This method was modified from Innis *et al.* (1990). A standard PCR reaction consisted of ~25 ng of template DNA, 1 µM of a pair of primers, 200 µM dNTPs, 1-2 U of Taq DNA polymerase, and 1×PCR reaction buffer (Table 4.2) in a 25 µl reaction. PCR cycling conditions were as follows: Initial denaturation at 95°C for 5 min, followed by (30 s at 95°C, 30 s at the melting temperature of the primers, 1 min per 1 kb of product at 72°C)^{30}cycles. Then, a final elongation at 72°C for 10 min and a cooling down of the reaction to 4°C followed. All PCR reactions were performed using a TPERSONAL Thermocycler (BIOMETRA). PCR reactions with Phusion™ High-Fidelity or Pfu DNA polymerase were performed according to the manufacturers protocol.

5.4.5 Quantitative real-time PCR (qPCR)

To remove unwanted genomic DNA, all RNA samples were treated with DNaseI Amplification Grade (INVITROGEN) as described by the supplier. Next, first-strand cDNA synthesis was carried out using SuperScript™II Reverse Transcriptase (INVITROGEN) from 1 µg of total RNA as described in the manufacturers protocol. qPCR was performed using an iCycler MyiQ™ (BIORAD) along with MESA GREEN qPCR MasterMix Plus for SYBR® Assay with fluorescein (EUROGENTEC). Cycling conditions were as follows: Initial denaturation at 95°C for 3 min, followed by 40 cycles of (95°C for 30 sec, 56°C for 30 sec, 72°C for 30 sec).

Each qPCR reaction contained 12.5 µl of MESA GREEN, 0.5 µl of each primer (10 µM), 10.5 µl of H_2O_{DEPC}, and 1 µl of a cDNA template. All cDNA samples were used in 1:5 dilutions and all samples were tested in three biological replicates including three technical replicates each. At the end of each qPCR, a melt curve analysis was carried out by rising the temperature from 57°C to 97°C to check for unspecific PCR products or primer dimers formed during the reaction.

The relative expression was calculated by normalizing the threshold cycle values (Ct) of the target gene against the Ct values of the housekeeping gene as indicated in the following equation:

$$\text{Relative Expression} = 2^{(\text{Ct}_{TEF} - Ct_{Target\ Gene})}$$

As houskeeping gene for *M. truncatula* and *R. irregularis*, the trans-elongation factor 1α, *TEF*, (*M. truncatula*: TC106470, *R. irregularis*: DQ282611) was used. Standard deviations were calculated by means of three biological replicates. Oligonucleotides used for qPCR are listed in Table 4.4.

5.4.6 Site-directed mutagenesis

For deletion or replacement of defined nucleotides in vectors, the QuikChange® site-directed mutagenesis kit (STRATAGENE) was used as described by the supplier. Therefore, mutated primers were designed with primerX[4] using the primer design protocol for QuikChange®. The respective primer sequences are listed in Table 4.3.

[4]www.bioinformatics.org/primerx/cgi-bin/DNA_1.cgi

5.4.7 RNAi-mediated gene silencing and overexpression studies

RNAi-mediated silencing (Fire *et al.*, 1998) and overexpression constructs were cloned as described in Section 4.3.2.

For each RNAi and overexpression construct, 70 composite *M. truncatula* plants were screened for transformed roots, of which 10 independent transformed roots were excised and propagated, resulting in a hairy root line. The rate of silencing or overexpression of each line was tested by qPCR and the best lines were selected for mycorrhization assays (hairy root lines are listed in Table 4.6).

Results of one biological replicate were obtained by the following procedure: 15 two-compartment mycorrhization plates were prepared for each construct and harvested at 17 dpi. Then, all roots of the same construct were pooled and divided into three parts used for RNA extraction, for ink staining (quantification of mycorrhizal colonization), and for WGA-Fluorescein staining (morphological analysis of AMS).

5.5 Biochemical methods

5.5.1 Yeast two-hybrid system

The yeast two-hybrid system is a commonly used method to study protein-protein interactions (Fields & Song, 1989). For the experiments of this work, the Gal4 based yeast-two hybrid system (Chien *et al.*, 1991) according to the Matchmaker™ Library Construction & Sreening Kit (Clontech) has been used. In this assay, the modular Gal4 transcription factor is split into the DNA-binding (DBD) and the transactivation domain (TAD), whereas each is fused to a protein of interest. If the two proteins interact, the Gal4 transcription factor can bind to the upstream activating sequence

(UAS) and activate the transcription of reporter genes. The yeast strain AH109 carries the four reporter genes *ADE2*, *HIS3*, *lacZ* and *MEL1* under control of the Gal4 UAS. *ADE2* and *HIS3* are nutritional selection markers, whereas, *MEL1* and *lacZ* encode for α- and β-galactosidase respectively, that can be used for colorimetric assays. In this work, positive interactions were screened by the growth ability of co-transformants on either SD-LWH (low stringency) or on SD-LWHA (high stringency) selection media. Furthermore, the growth of all co-transformants was assayed on SD-LW medium to indicate the presence of bait and prey vectors.

5.5.1.1 Direct yeast two-hybrid tests

Direct yeast two-hybrid tests were carried out by co-transformation of the AH109 strain with the respective bait and prey plasmids. Transformed colonies growing on SD-LW selection plates (Section 4.5.2) at 30°C were picked 3–5 days after transformation, streaked out again on SD-LW plates and incubated for another 3–5 days to increase biomass. Then, at least ten colonies per co-transformation (biological replicates) were replica streaked on SD-LW, and SD-LWHA plates to test for protein–protein interaction.

To visualize the strength of protein-protein interaction, a spotting assay was used. Therefore, two yeast colonies per co-transformation (biological replicates) were resuspended each in 500 µl of SD-LWHA liquid medium. After well vortexing, 10-fold dilution series of up to 10^{-3} were made and 5 µl of each dilution was spotted on SD-LW and SD-LWH or SD-LWHA plates, respectively. Each spotting plate included the two controls p53–T (positive protein interaction, Li & Fields (1993)) and Lam–T (negative interaction; plasmids are listed in Section 4.3.1). Plates were incubated at 30°C for 5 days.

5.5.2 Yeast protein extraction

Yeast protein extracts were prepared according to the yeast protocols handbook (Clontech) using the Urea/SDS method. Therefore, 50 ml of the respective SD selection medium were inoculated with 5 ml of a fresh grown overnight culture and incubated at 30°C, 200 rpm to a final OD_{600} of 0.6. Then, cultures were transferred in 50 ml centrifugation tubes halfway filled with ice and centrifuged for 5 min at 1,500 rpm at 4°C. The supernatant was discarded, the pellet resuspended in 25 ml of ice cold $_{d}H_2O$ and centrifuged for 5 min at 1,500 g at 4°C. The supernatant was discarded, tubes immediately frozen in liquid nitrogen and stored at −80°C until further use.

For protein extraction, frozen cell pellets were thawed on ice, and resuspended each in 500 µl of freshly prepared pre-warmed (70°C) complete cracking buffer (10 ml of cracking buffer (see Table 4.2), 500 µl of PMSF solution (see Table 4.1), 100 µl ß-mercaptoethanol). Then, the suspension was transferred into 2 ml centrifugation tubes containing 500 µl of glass beads (250 µm diameter, Sigma). Samples were incubated at 70°C for 2 min and chilled on ice for 1 min. Next, samples were vortexed for at least 1 min, and centrifuged at 14,000 rpm, 4°C for 5 min. The supernatant was transferred into a 1.5 ml centrifugation tube and incubated at 95°C for 5 min. Samples were chilled on ice for 1 min and centrifuged at maximum rpm for 1 min. The supernatant was immediately loaded on a polyacrylamide gel or stored at −20°C for later use.

5.5.3 Protein extraction of *N. benthamiana* leaves

Protein preparation from infiltrated tobacco leaves was carried out as described in Schütze *et al.* (2009). For each construct, 5 leaf discs (1 cm diameter) were excised, transferred into a 1.5 ml centrifugation tube, and homogenized under liquid nitrogen with a plastic pestle (Rotilabo®micro-homogenizers, ROTH). Then, 300 µl of hot 1×Laemmli buffer (see Table 4.2) were added, and incubated at 95°C for 5 min. Samples were chilled on ice and centrifuged for 5 min at 13,000 rpm to remove debris. The supernatant was loaded on a polyacrylamide gel or stored at −20°C.

5.5.4 Recombinant protein expression in *E. coli* and cell disruption

For pull-down assays, recombinant proteins were expressed in the *E. coli* BL21(DE) pLysS strain as follows. LB medium supplied with the appropriate antibiotics was inoculated 1:100 with a well-grown overnight culture. The culture was incubated at 37°C, 180 rpm until OD_{600} of 0.8. Then, 1 ml of the culture was transferred into a 1.5 ml centrifugation tube, centrifuged at 13,000 rpm for 1 min and the pellet frozen at −20°C (non-induced control). Induction was carried out by adding 1:1,000 of a 0.1 M (w/v) IPTG (Isopropyl β-D-1-thiogalactopyranoside) solution and protein expression occurred at 21°C, 180 rpm for 6 hours. As induction control, 1 ml of the culture was removed and proceeded as described above. The remaining culture was then transferred into 250 ml centrifugation bottles and centrifuged for 10 min, 5,000 rpm at 4°C. The supernatant was discarded, the cell pellet immediately frozen in liquid nitrogen and stored at −80°C until further use or immediately used for cell disruption.

Therefore, the *E. coli* cell pellets were resuspended in an adequate 1×buffer depending on the respective downstream processing, and cell disruption was carried out using the EmulsiFlex®C3 homogenizer (AVENTIN) as described by the supplier. The cell lysate was transferred into 20 ml centrifugation tubes and centrifuged at 20,000 rpm at 4°C for 10 min in a Beckman J2-21 (BECKMAN COULTER) centrifuge. The supernatant containing the soluble proteins (crude extract) was used for the *in vitro* pull-down assay.

5.5.5 GST pull-down assay

To confirm the positive yeast two-hybrid interactions between KPI106, KPI104 and CP, a GST pull-down assay was carried out. Therefore, resin-bound glutathione (GST Bind™ · Resin; NOVAGEN) was used to immobilize the GST-CP bait and to pull down the prey proteins 6×His-KPI106 and 6×His-KPI104 out of an *E. coli* crude extract. It has to be noticed here that the GST-CP construct includes the pro-domain of the cysteine protease. It was reported by Broemme *et al.* (2004) that the presence of pro-domains is essential for correct folding of papain-like cysteine proteases when heterologously expressed in *E. coli*. The respective plasmids used for recombinant expression of 6×His-KPI106, 6×His-KPI104 and GST-CP are listed in Section 4.3.2.

Pull-down experiment: 600 µl of resin-bound glutathione were centrifuged at 1,500 rpm, 4°C for 10 min, washed with 1×GST buffer (Table 4.2) and after another centrifugation step resuspended in 150 µl of 1×GST buffer. Then, 15 ml of a freshly prepared GST-CP crude extract (out of 500 ml cell culture) were gently mixed with the pre-washed resin-bound glutathione and incubated in a 15 ml centrifugation tube, rotating overnight at 4°C. The next day, the resin suspension was centrifuged at 1,500 rpm, 4°C for 10 min. Next, 1 ml of the supernatant was frozen at –20°C as flow-through control 1 (F1; Figure S3), and the resin pellet was washed two times, each with 1.5 ml of 1×GST buffer. After each wash, 500 µl of each supernatant were frozen for later analysis (W1, W2).

Meanwhile, the crude extracts of 6×His-KPI106 and 6×His-KPI104 (each from 250 ml of cell culture) were prepared as described in Section 5.5.4, transferred into 15 ml centrifugation tubes and each crude extract was gently mixed with 750 µl of the resin-bound GST-CP. Incubation occurred at 4°C, rotating for 5 hours. Then, samples were centrifuged at 1,500 rpm, 4°C for 10 min and 500 µl of each supernatant was frozen as flow-through control 2 (F2; Figure S3). Samples were washed two times, each with 1.5 ml of 1×GST buffer (W3, W4). After the second wash, the pellets were resuspended in 150 µl of 1×GST elution buffer (Table 4.2) and incubated at 4°C, rotating for 10 min to elute the pull-down complex from the resin. The samples were centrifuged at 4°C, 1,500 rpm, 5 min and the supernatant was carefully transferred to a new 1.5 ml centrifugation tube (Elution 1; Figure S3). A second and a third elution followed and samples were stored at −20°C.

In addition, control pull-downs including free GST as bait (to rule out prey interactions with the GST-tag) and a bait-free control to confirm that the prey proteins do not bind to the immobilized glutathione were conducted accordingly.

All samples collected during the pull-down experiments were loaded on an SDS gel and analyzed *via* western blot.

5.5.6 SDS-PAGE and western blot analysis

Protein separation was carried out based on the discontinuous SDS-PAGE (sodium dodecyl sulfate polyacrylamide gel electrophoresis) system of Laemmli (1970). Therefore, standard separating gels with a concentration of 12 % of acrylamide were used (Table 5.1). Electrophoresis was performed in 1×Tris-glycine buffer (Table 4.2) in a Mini-PROTEAN® Tetra Cell (BioRad) system at 60 V until the dye front ran into the separating gel and then increased up to 120 V for protein separation. If not mentioned differently, protein samples were mixed with 5×Laemmli buffer (Table 4.2) and PageRuler Prestained Protein Ladder (Thermo Scientific) was used as standard.

To visualize separated proteins, polyacrylamide gels were incubated for 30 min in Coomassie staining solution (Table 4.1). After incubation with Coomassie destaining solution for 1 hour, protein bands were visible.

Table 5.1: Ingredients of 12 % polyacrylamide gels

Stock solution	Separating gel	Stacking gel
$_{d}H_2O$	2.2 ml	2.2 ml
1.5 M Tris-HCl (pH 8.8)	1.5 ml	–
1 M Tris-HCl (pH 6.8)	–	375 µl
40 % (w/v) acrylamide	1.5 ml	375 µl
10 % (w/v) SDS	50 µl	30 µl
10 % (w/v) APS	50 µl	30 µl
TEMED	2.5 µl	3 µl

For western blot analysis, proteins were transferred to a Protran® Nitrocellulose membrane (WHATMAN) using the Mini Trans-Blot Module (BIORAD). The transfer was performed at 30 V at 4°C overnight in 1×blotting buffer (Table 4.2). For detection of proteins containing 6×His-, GST-, or GFP-tags, the membrane was blocked in PBST containing 10 % (w/v) non-fat milk powder for 1 hour at RT. Then, incubation with the primary antibody (mouse α-His 1:3000; mouse α-GST 1:6000; rabbit α-GFP 1:4000, all SIGMA) occurred in PBST containing 5 % (w/v) of non-fat milk powder at RT for 1 hour. For detection of proteins containing c-myc-tag, the membrane was blocked in PBS containing 5 % (w/v) of BSA and incubation with the primary antibody (mouse α-c-myc 1:4500 SIGMA) occurred in PBS with 2.5 % (w/v) of BSA at RT. In all cases, the membrane was washed three times with PBST and incubation with a secondary antibody HRP-conjugate 1:8000 (rabbit anti-mouse IgG-HRP; goat anti-rabbit IgG-HRP) occurred for 1 hour. After three washes with PBST, the membrane was transferred to a plastic foil and the solutions for chemiluminescence detection (1 ml of solution A, 100 µl of solution B, 0.5 µl H_2O_2) were added. The foil was sealed, incubated for 2 min and detection was performed using a CHEMI-SMART-5100 (PEQLAB), supplied with the software Chemi-Capt (PEQLAB) according to the manufacturers instructions.

5.6 *In silico* analyses

5.6.1 T-test

Statistical significance of the quantified mycorrhizal colonization parameters was tested by performance of an independent one-tailed Student´s t-test in MICROSOFT Excel. Statistical significance was assumed if the null hypothesis was rejected at the 0.01 or 0.05 level, indicated by ** or *, respectively, in the according figure legends.

5.6.2 Sequencing and sequence analyses

All DNA sequencing reactions were performed by EUROFINS MWG OPERON. For evaluation of the nucleotide sequences, the ApE software[5] was used.

Protein sequences were analyzed using the servers SignalP4.1[6] (Petersen *et al.*, 2011) and TargetP1.1[7] (Emanuelsson *et al.*, 2000) for prediction of putative secretion signal peptides. For prediction of specific protein domains, the amino acid sequences were blasted against the pfam database[8] (Punta *et al.*, 2012), the MEROPS Peptidase database[9](Rawlings *et al.*, 2012) and the non-redundant (nr) protein sequences database of NCBI[10] (Altschul *et al.*, 1990). The presented blast results are based on queries carried out in July 2013.

[5]biologylabs.utah.edu/jorgensen/wayned/ape/
[6]www.cbs.dtu.dk/services/SignalP/
[7]http://www.cbs.dtu.dk/services/TargetP/
[8]pfam.sanger.ac.uk/
[9]merops.sanger.ac.uk/inhibitors/
[10]blast.ncbi.nlm.nih.gov/Blast.cgi

5.6.3 Phylogenetic analyses

Phylogenetic trees are based on cDNA sequences obtained from the *Medicago* hapmap3.5 database[11] or on protein sequences obtained from NCBI. *In silico* expression analyses were carried out using the Gene expression Atlas[12] (He *et al.*, 2009), and the DFCI *M. truncatula* Gene Index[13] as EST database. The presented results are based on queries carried out in July 2013. Alignments were created with CLUSTALX[14] (Larkin *et al.*, 2007), and the respective trees by the MEGA5.0 software[15] (Tamura *et al.*, 2011) using the Neighbor-Joining algorithm and a bootstrap value of 1000.

5.6.4 Three dimensional protein models

The web server PHYRE2[16] (Kelley & Sternberg, 2009) predicts protein structures based on homology modeling, and was used to construct the models for SCP1 and KPI106. In homology modeling, the structure of a target protein (atomic-resolution model) is predicted based on its amino acid sequence in relation to an already known structure of a homologous protein. Therefore, the first step is to identify a suitable template structure of the target protein by PSI-blast (Altschul *et al.*, 1997). The PHYRE^2 output for SCP1 revealed the carboxypeptidase Y from *S. cerevisiae* (RCSB PDB entry: 1wpxA1), and in case of KPI106 the protease inhibitor API-A from S. *sagittifolia* (RCSB PDB entry: 3E8L_C) as best model templates. The API-A protein (Bao *et al.*, 2009) contains three disulfide bonds as does KPI106. The KPI106 model was adjusted to API-A using SWISS-MODEL[17] (Schwede *et al.*, 2003). For both, the SCP1 and

[11] www.medicagohapmap.org/
[12] mtgea.noble.org/v2/
[13] compbio.dfci.harvard.edu/tgi/cgi-bin/tgi/gimain.pl?gudb=medicago
[14] www.clustal.org/download/current/
[15] www.megasoftware.net/
[16] www.sbg.bio.ic.ac.uk/phyre2/html/page.cgi?id=index
[17] serverwww.expasy.org/structural_bioinformatics

the KPI106 model, Ramachandran plots[18] (Furnham *et al.*, 2008) as well as z-score and energy plots[19] (Wiederstein & Sippl, 2007) were applied to evaluate the quality of the protein models and results can be reviewed in Heidt, 2012 (KIT).

5.6.5 *In silico* docking

In silico docking was performed by the docking server PATCHDOCK[20] (Schneidman-Duhovny *et al.*, 2005) using a geometry based molecular docking algorithm. With the parameters 4.0 as clustering root mean squared deviation and the selected complex type "Enzyme-Inhibitor", a docking model of KPI106 and SCP1 was generated. The docking model with the highest score of 13964, an area of 2152.3 and an atomic contact energy of 300.43 was used for further analysis and visualized with UCSF chimera[21] (Pettersen *et al.*, 2004).

[18] mordred.bioc. cam.ac.uk/~rapper/rampage.php
[19] prosa.services.came.sbg.ac.at/prosa.php
[20] bioinfo3d.cs.tau.ac.il/PatchDock/
[21] www.cgl.ucsf.edu/chimera/

REFERENCES

Ahmed, S. U., Rojo, E., Kovaleva, V., Venkataraman, S., Dombrowski, J. E., Matsuoka, K., & Raikhel, N. V. (2000). The plant vacuolar sorting receptor atelp is involved in transport of nh(2)-terminal propeptide-containing vacuolar proteins in arabidopsis thaliana. *J. Cell Biol.* **149** (7), 1335–1344.

Akiyama, K., Matsuzaki, K., & Hayashi, H. (2005). Plant sesquiterpenes induce hyphal branching in arbuscular mycorrhizal fungi. *Nature,* **435** (7043), 824–827.

Alexander, T., Toth, R., Meier, R., & Weber, H. C. (1989). Dynamics of arbuscule development and degeneration in onion, bean, and tomato with reference to vesicular–arbuscular mycorrhizae in grasses. *Can. J. Bot.* **67** (8), 2505–2513.

Altschul, S., Madden, T., Schäffer, A., Zhang, J., Zhang, Z., Miller, W., & Lipman, D. (1997). Gapped blast and psi-blast: a new generation of protein database search programs. *Nucleic Acids Res.* **25** (17), 3389–3402.

Altschul, S., Warren, G., Miller, W., Myers, E., & Lipman, D. (1990). Basic local alignment search tool. *J. Mol. Biol.* **215** (3), 403–410.

Ané, J.-M., Kiss, G. B., Riely, B. K., Penmetsa, R. V., Oldroyd, G. E. D., Ayax, C., Lévy, J., Debellé, F., Baek, J.-M., Kalo, P., Rosenberg, C., Roe, B. A., Long, S. R., Dénarié, J., & Cook, D. R. (2004). Medicago truncatula dmi1 required for bacterial and fungal symbioses in legumes. *Science,* **303** (5662), 1364–1367.

Ausubel, F., Brent, R., Kingston, R., Moore, D., Seidman, J., Smith, J., & Struhl, K. (1999). *Short protocols in molecular biology: a compendium of methods from current protocols in molecular biology.* New York: John Wiley.

Bago, B., Pfeffer, P. E., & Shachar-Hill, Y. (2000). Carbon metabolism and transport in arbuscular mycorrhizas. *Plant Physiol.* **124** (3), 949–958.

Baier, M. C., Keck, M., Gödde, V., Niehaus, K., Küster, H., & Hohnjec, N. (2010). Knockdown of the symbiotic sucrose synthase mtsucs1 affects arbuscule maturation and maintenance in mycorrhizal roots of medicago truncatula. *Plant Physiol,* **152** (2), 1000–1014.

Bais, H., Weir, T., Perry, L., Gilroy, S., & Vivanco, J. (2006). The role of root exudates in rhizosphere interactions with plants and other organisms. *Annu. Rev. Plant Biol.* **57** (1), 233–266. PMID: 16669762.

Baker, D., Shiau, A. K., & Agard, D. A. (1993). The role of pro regions in protein folding. *Curr. Opin. Cell Biol.* **5** (6), 966–970.

Bao, R., Zhou, C. Z., Jiang, C., Lin, S. X., Chi, C. W., & Chen, Y. (2009). The ternary structure of the double-headed arrowhead protease inhibitor api-a complexed with two trypsins reveals a novel reactive site conformation. *J Biol Chem,* **284** (39), 26676–26684.

Barrett, A. J., Rawlings, N. D., & O'Brien, E. A. (2001). The merops database as a protease information system. *J. Struct. Biol.* **134** (2–3), 95–102.

Bayrhuber, M., Vijayan, V., Ferber, M., Graf, R., Korukottu, J., Imperial, J., Garrett, J. E., Olivera, B. M., Terlau, H., Zweckstetter, M., & Becker, S. (2005). Conkunitzin-s1 is the first member of a new kunitz-type neurotoxin family: Structural and functional characterization. *J. Biol. Chem.* **280** (25), 23766–23770.

Bécard, G. & Fortin, J. A. (1988). Early events of vesicular-arbuscular mycorrhiza formation on ri t-dna early events of vesicular-arbuscular mycorrhiza formation on ri t-dna transformed roots. *New. Phytol.* **108** (2), 211–218.

Berezniuk, I., Vu, H. T., Lyons, P. J., Sironi, J. J., Xiao, H., Burd, B., Setou, M., Angeletti, R. H., Ikegami, K., & Fricker, L. D. (2012). Cytosolic carboxypeptidase 1 is involved in processing α- and β-tubulin. *J. Biol. Chem.* **287** (9), 6503–6517.

Besserer, A., Puech-Pagès, V., Kiefer, P., Gomez-Roldan, V., Jauneau, A., Roy, S., Portais, J. C., Roux, C., Bécard, G., & Séjalon-Delmas, N. (2006). Strigolactones stimulate arbuscular mycorrhizal fungi by activating mitochondria. *PLoS Biol,* **4** (7), e226.

Bethke, P. C., Schuurink, R., & Jones, R. L. (1997). Hormonal signalling in cereal aleurone. *J. Exp. Botany,* **48** (7), 1337–1356.

Bewley, J. D. (1997). Seed germination and dormancy. *Plant Cell,* **9** (7), 1055–1066.

Birk, Y. (2003). *Plant protease inhibitors: Significance in nutrition, plant protection, cancer prevention and genetic engineering.* Springer Berlin Heidelberg.

Blow, D. M. & Sweet, R. M. (1974). Mode of action of soybean trypsin inhibitor (kunitz) as a model for specific protein-protein interactions. *Nature,* **249** (5452), 54–74.

Bode, W. & Huber, R. (1992). Natural protein proteinase inhibitors and their interaction with proteinases. *Euro. J. Biochem.* **204** (2), 433–451.

Boisson-Dernier, A., Chabaud, M., Garcia, F., Bécard, G., Rosenberg, C., & Barker, D. G. (2001). Agrobacterium rhizogenes-transformed roots of medicago truncatula for the study of nitrogen-fixing and endomycorrhizal symbiotic associations. *Mol. Plant Microbe Interact.* **14** (6), 695–700.

Boller, T. (1995). Chemoperception of microbial signals in plant cells. *Annu.Rev. Plant Physiol.* **46** (1), 189–214.

Boller, T. & Felix, G. (2009). A renaissance of elicitors: Perception of microbe-associated molecular patterns and danger signals by pattern-recognition receptors. *Annu. Rev. Plant Biol.* **60** (1), 379–406. PMID: 19400727.

Bonfante, P. & Perotto, S. (1995). Strategies of arbuscular mycorrhizal fungi when infecting host plants. *New Phytol.* **130** (1), 3–21.

Bonfante, P. & Requena, N. (2011). Dating in the dark: how roots respond to fungal signals to establish arbuscular mycorrhizal symbiosis. *Curr Opin Plant Biol,* **14** (4), 451–457.

Bonfante-Fasolo, P. (1984). Va mycorrhizae. chapter Anatomy and morphology of VA Mycorrhizae, pp. 5–33. CRC Press Boca Raton, Florida.

Bonfante-Fasolo, P., Faccio, A., Perotto, S., & Schubert, A. (1990). Correlation between chitin distribution and cell wall morphology in the mycorrhizal fungus glomus versiforme. *Mycol. Res.* **94** (2), 157–165.

Bradshaw, H., Ceulemans, R., Davis, J., & Stettler, R. (2000). Emerging model systems in plant biology: Poplar (populus) as a model forest tree. *J. Plant Growth Reg.* **19** (3), 306–313.

Breuillin, F., Schramm, J., Hajirezaei, M., Ahkami, A., Favre, P., Druege, U., Hause, B., Bucher, M., Kretzschmar, T., Bossolini, E., Kuhlemeier, C., Martinoia, E., Franken, P., Scholz, U., & Reinhardt, D. (2010). Phosphate systemically inhibits development of arbuscular mycorrhiza in petunia hybrida and represses genes involved in mycorrhizal functioning. *Plant J.* **64** (6), 1002–1017.

Broemme, D., Nallaseth, F. S., & Turk, B. (2004). Production and activation of recombinant papain-like cysteine proteases. *Methods,* **32** (2), 199–206.

Broghammer, A., Krusell, L., Blaise, M., Sauer, J., Sullivan, J. T., Maolanon, N., Vinther, M., Lorentzen, A., Madsen, E. B., Jensen, K. J., Roepstorff, P., Thirup, S., Ronson, C. W., Thygesen, M. B., & Stougaard, J. (2012). Legume receptors perceive the rhizobial lipochitin oligosaccharide signal molecules by direct binding. *Proc. Natl. Acad. Sci.* **109** (34), 13859–13864.

Catoira, R., Galera, C., de Billy, F., Penmetsa, R. V., Journet, E.-P., Maillet, F., Rosenberg, C., Cook, D., Gough, C., & Dénarié, J. (2000). Four genes of medicago truncatula controlling components of a nod factor transduction pathway. *Plant Cell,* **12** (9), 1647–1665.

Cercos, M., Urbez, C., & Carbonell, J. (2003). A serine carboxypeptidase gene (pscp), expressed in early steps of reproductive and vegetative development in pisum sativum, is induced by gibberellins. *Plant Mol. Biol.* **51** (2), 165–174.

Chabaud, M., Genre, A., Sieberer, B. J., Faccio, A., Fournier, J., Novero, M., Barker, D. G., & Bonfante, P. (2011). Arbuscular mycorrhizal hyphopodia and germinated spore exudates trigger ca2+ spiking in the legume and nonlegume root epidermis. *New Phytol.* **189** (1), 347–355.

Charpentier, M., Vaz Martins, T., Granqvist, E., Oldroyd, G. E. D., & Morris, R. J. (2013). The role of dmi1 in establishing ca^{2+} oscillations in legume symbioses. *Plant Signal. Behav.* **8** (2), e22894.

Chien, C. T., Bartel, P. L., Sternglanz, R., & Fields, S. (1991). The two-hybrid system: a method to identify and clone genes for proteins that interact with a protein of interest. *Proceedings of the National Academy of Sciences,* **88** (21), 9578–9582.

Clouse, S. D., Langford, M., & McMorris, T. C. (1996). A brassinosteroid-insensitive mutant in arabidopsis thaliana exhibits multiple defects in growth and development. *Plant Physiol.* **111** (3), 671–678.

Colbert, T., Till, B. J., Tompa, R., Reynolds, S., Steine, M. N., Yeung, A. T., McCallum, C. M., Comai, L., & Henikoff, S. (2001). High-throughput screening for induced point mutations. *Plant Physiol,* **126** (2), 480–484.

Cook, D. (1999). Medicago truncatula — a model in the making!: Commentary. *Current Opinion in Plant Biology,* **2** (4), 301–304.

Cox, G. & Sanders, F. (1974). Ultrastructure of the host-fungus interface in a vesicular-arbuscular mycorrhiza. *New Phytologist,* **73** (5), 901–912.

Criekinge, W. & Beyaert, R. (1999). Yeast two-hybrid: State of the art. *Biol. Proced. Online,* **2** (1), 1–38.

de Souza Cândido, E., Pinto, M. F., Pelegrini, P. B., Lima, T. B., Silva, O. N., Pogue, R., Grossi-de Sá, M. F., & Franco, O. L. (2011). Plant storage proteins with antimicrobial activity: novel insights into plant defense mechanisms. *FASEB J,* **25** (10), 3290–3305.

Delaux, P.-M., Bécard, G., & Combier, J.-P. (2013). Nsp1 is a component of the myc signaling pathway. *New Phytol.* **199** (1), 59–65.

Demchenko, K., Winzer, T., Stougaard, J., Parniske, M., & Pawlowski, K. (2004). Distinct roles of lotus japonicus symrk and sym15 in root colonization and arbuscule formation. *New Phytol.* **163** (2), 381–392.

Denarie, J., Debelle, F., & Prome, J.-C. (1996). Rhizobium lipo-chitooligosaccharide nodulation factors: Signaling molecules mediating recognition and morphogenesis. *Annu. Rev. Biochem.* **65** (1), 503–535. PMID: 8811188.

Dodson, G. & Wlodawer, A. (1998). Catalytic triads and their relatives. *Trends Biochem. Sci.* **23** (9), 347–352.

Dominguez, F. & Cejudo, F. J. (1999). Patterns of starchy endosperm acidification and protease gene expression in wheat grains following germination. *Plant Physiol.* **119** (1), 81–88.

Drag, M. & Salvesen, G. S. (2010). Emerging principles in protease-based drug discovery. *Nat. Rev. Drug Discov.* **9** (9), 690–701.

Drissner, D., Kunze, G., Callewaert, N., Gehring, P., Tamasloukht, M., Boller, T., Felix, G., Amrhein, N., & Bucher, M. (2007). Lyso-phosphatidylcholine is a signal in the arbuscular mycorrhizal symbiosis. *Science,* **318** (5848), 265–268.

Dunn, R. (2001). Determination of protease mechanism. In: *Proteolytoc enzymes: a practical approach,* (Beynon, R. & Bond, J. S., eds) 2nd edition. Oxford: Univ. Press.

Dy, C. Y., Buczek, P., Imperial, J. S., Bulaj, G., & Horvath, M. P. (2006). Structure of conkunitzin-s1, a neurotoxin and kunitz-fold disulfide variant from cone snail. *Acta Crystallo.* **62** (9), 980–990.

Emanuelsson, O., Nielsen, H., Brunak, S., & von Heijne, G. (2000). Predicting subcellular localization of proteins based on their n-terminal amino acid sequence. *J Mol Biol,* **300** (4), 1005–1016.

Endre, G., Kereszt, A., Kevei, Z., Mihacea, S., Kalo, P., & Kiss, G. B. (2002). A receptor kinase gene regulating symbiotic nodule development. *Nature,* **417** (6892), 962–966.

Fan, X., Olson, S. J., Blevins, L. S., Allen, G. S., & Johnson, M. D. (2002). Immunohistochemical localization of carboxypeptidases d, e, and z in pituitary adenomas and normal human pituitary. *J. Histochem. Cytochem.* **50** (11), 1509–1515.

Feddermann, N. & Reinhardt, D. (2011). Conserved residues in the ankyrin domain of vapyrin indicate potential protein-protein interaction surfaces. *Plant Signal. Behav.* **6** (5), 680–684.

Fernandez-Gonzalez, A., Spada, A. R. L., Treadaway, J., Higdon, J. C., Harris, B. S., Sidman, R. L., Morgan, J. I., & Zuo, J. (2002). Purkinje cell degeneration (pcd) phenotypes caused by mutations in the axotomy-induced gene, nna1. *Science,* **295** (5561), 1904–1906.

Fields, S. (1993). The two-hybrid system to detect protein-protein interactions. *Methods,* **5** (2), 116–124.

Fields, S. & Song, O. (1989). A novel genetic system to detect protein–protein interactions. *Nature,* **340** (6230), 245–246.

Finlay, R. D. (2008). Ecological aspects of mycorrhizal symbiosis: with special emphasis on the functional diversity of interactions involving the extraradical mycelium. *J Exp Bot,* **59** (5), 1115–1126.

Fire, A., Xu, S., Montgomery, M. K., Kostas, S. A., Driver, S. E., & Mello, C. C. (1998). Potent and specific genetic interference by double-stranded rna in caenorhabditis elegans. *Nature,* **391** (6669), 806–811.

Fliegmann, J., Canova, S., Lachaud, C., Uhlenbroich, S., Gasciolli, V., Pichereaux, C., Rossignol, M., Rosenberg, C., Cumener, M., Pitorre, D., Lefebvre, B., Gough, C., Samain, E., Fort, S., Driguez, H., Vauzeilles, B., Beau, J.-M., Nurisso, A., Imberty, A., Cullimore, J., & Bono, J.-J. (2013). Lipo-chitooligosaccharidic symbiotic signals are recognized by lysm receptor-like kinase lyr3 in the legume medicago truncatula. *ACS Chemical Biology,* **8** (9), 1900–1906.

Floss, D. S., Hause, B., Lange, P. R., Küster, H., Strack, D., & Walter, M. H. (2008a). Knock-down of the mep pathway isogene 1-deoxy-d-xylulose 5-phosphate synthase 2 inhibits formation of arbuscular mycorrhiza-induced apocarotenoids, and abolishes normal expression of mycorrhiza-specific plant marker genes. *Plant J,* **56** (1), 86–100.

Floss, D. S., Schliemann, W., Schmidt, J., Strack, D., & Walter, M. H. (2008b). Rna interference-mediated repression of mtccd1 in mycorrhizal roots of medicago truncatula causes accumulation of c27 apocarotenoids, shedding light on the functional role of ccd1. *Plant Physiol.* **148** (3), 1267–1282.

Foo, E., Ross, J. J., Jones, W. T., & Reid, J. B. (2013). Plant hormones in arbuscular mycorrhizal symbioses: an emerging role for gibberellins. *Ann. Bot.* **111**, 769–779.

Frank, B. (1885). Über die auf wurzelsymbiose beruhende ernährung gewisser bäume durch unterirdische pilze. *Ber. Deutsch. Bot. Ges.* **3**, 128–145.

Frank, B. A. (2005). On the nutritional dependence of certain trees on root symbiosis with belowground fungi (an english translation of a.b. frank's classic paper of 1885). *Mycorrhiza,* **15**, 267–275.

Fry, B. G., Roelants, K., Champagne, D. E., Scheib, H., Tyndall, J. D., King, G. F., Nevalainen, T. J., Norman, J. A., Lewis, R. J., Norton, R. S., Renjifo, C., & de la Vega, R. C. R. (2009). The toxicogenomic multiverse: Convergent recruitment of proteins into animal venoms. *Annu. Rev. Genom. Hum. Genet.* **10** (1), 483–511. PMID: 19640225.

Furnham, N., de Bakker, P., Gore, S., Burke, D., & Blundell, T. (2008). Comparative modelling by restraint-based conformational sampling. *BMC Struc. Biol.* **8** (1), 7.

Garcia-Lorenzo, M., Sjodin, A., Jansson, S., & Funk, C. (2006). Protease gene families in populus and arabidopsis. *BMC Plant Biol.* **6** (1), 30.

Gaude, N., Bortfeld, S., Duensing, N., Lohse, M., & Krajinski, F. (2012). Arbuscule-containing and non-colonized cortical cells of mycorrhizal roots undergo extensive and specific reprogramming during arbuscular mycorrhizal development. *The Plant Journal,* **69** (3), 510–528.

Genre, A., Chabaud, M., Balzergue, C., Puech-Pagès, V., Novero, M., Rey, T., Fournier, J., Rochange, S., Bécard, G., Bonfante, P., & Barker, D. G. (2013). Short-chain chitin oligomers from arbuscular mycorrhizal fungi trigger nuclear ca2+ spiking in medicago truncatula roots and their production is enhanced by strigolactone. *New Phytol.* **198** (1), 190–202.

Genre, A., Chabaud, M., Faccio, A., Barker, D. G., & Bonfante, P. (2008). Prepenetration apparatus assembly precedes and predicts the colonization patterns of arbuscular mycorrhizal fungi within the root cortex of both medicago truncatula and daucus carota. *Plant Cell,* **20** (5), 1407–1420.

Genre, A., Chabaud, M., Timmers, T., Bonfante, P., & Barker, D. G. (2005). Arbuscular mycorrhizal fungi elicit a novel intracellular apparatus in medicago truncatula root epidermal cells before infection. *Plant Cell,* **17** (12), 3489–3499.

Genre, A., Ivanov, S., Fendrych, M., Faccio, A., Žárský, V., Bisseling, T., & Bonfante, P. (2012). Multiple exocytotic markers accumulate at the sites of perifungal membrane biogenesis in arbuscular mycorrhizas. *Plant Cell Physiol.* **53** (1), 244–255.

Giri, B., Kapoor, R., & Mukerji, K. (2007). Improved tolerance of acacia nilotica to salt stress by arbuscular mycorrhiza, glomus fasciculatum may be partly related to elevated k/na ratios in root and shoot tissues. *Microbial Ecology,* **54** (4), 753–760.

Gobbato, E., Marsh, J. F., Vernié, T., Wang, E., Maillet, F., Kim, J., Miller, J. B., Sun, J., Bano, S. A., Ratet, P., Mysore, K. S., Dénarié, J., Schultze, M., & Oldroyd, G. E. D. (2012). A gras-type transcription factor with a specific function in mycorrhizal signaling. *Current Biology,* **22** (23), 2236–2241.

Gomez, S. K., Javot, H., Deewatthanawong, P., Torres-Jerez, I., Tang, Y., Blancaflor, E. B., Udvardi, M. K., & Harrison, M. J. (2009). Medicago truncatula and glomus intraradices gene expression in cortical cells harboring arbuscules in the arbuscular mycorrhizal symbiosis. *BMC Plant Biol,* **9** (1), 1–19.

Gough, C. & Cullimore, J. (2011). Lipo-chitooligosaccharide signaling in endosymbiotic plant-microbe interactions. *Mol. Plant Microbe Interact.* **24** (8), 867–878.

Green, T. R. & Ryan, C. A. (1972). Wound-induced proteinase inhibitor in plant leaves: A possible defense mechanism against insects. *Science,* **18** (175), 776–777.

Groth, M., Kosuta, S., Gutjahr, C., Haage, K., Hardel, S. L., Schaub, M., Brachmann, A., Sato, S., Tabata, S., Findlay, K., Wang, T. L., & Parniske, M. (2013). Two lotus japonicus symbiosis mutants impaired at distinct steps of arbuscule development. *Plant J.* **75** (1), 117–129.

Groth, M., Takeda, N., Perry, J., Uchida, H., Dräxl, S., Brachmann, A., Sato, S., Tabata, S., Kawaguchi, M., Wang, T. L., & Parniske, M. (2010). Nena, a lotus japonicus homolog of sec13, is required for rhizodermal infection by arbuscular mycorrhiza fungi and rhizobia but dispensable for cortical endosymbiotic development. *Plant Cell,* **22** (7), 2509–2526.

Grudkowska, M. & Zagdanska, B. (2004). Multifunctional role of plant cysteine proteinases. *Acta Biochim. Pol.* **51** (3), 609–624.

Grunwald, U. (2004). *Analyse der Genregulation in Medicago truncatula und Pisum sativum während der Entwicklung arbuskulärer Mykorrhiza.* PhD thesis Philipps-Universität Marburg.

Grunwald, U., Nyamsuren, O., Tamasloukht, M., Lapopin, L., Becker, A., Mann, P., Gianinazzi-Pearson, V., Krajinski, F., & Franken, P. (2004). Identification of mycorrhiza-regulated genes with arbuscule development-related expression profile. *Plant Mol Biol,* **55** (4), 553–566.

Guether, M., Balestrini, R., Hannah, M., He, J., Udvardi, M. K., & Bonfante, P. (2009a). Genome-wide reprogramming of regulatory networks, transport, cell wall and membrane biogenesis during arbuscular mycorrhizal symbiosis in lotus japonicus. *New Phytol.* **182** (1), 200–212.

Guether, M., Neuhäuser, B., Balestrini, R., Dynowski, M., Ludewig, U., & Bonfante, P. (2009b). A mycorrhizal-specific ammonium transporter from lotus japonicus acquires nitrogen released by arbuscular mycorrhizal fungi. *Plant Physiol.* **150** (1), 73–83.

Guimil, S., Chang, H. S., Zhu, T., Sesma, A., Osbourn, A., Roux, C., Ioannidis, V., Oakeley, E. J., Docquier, M., Descombes, P., Briggs, S. P., & Paszkowski, U. (2005). Comparative transcriptomics of rice reveals an ancient pattern of response to microbial colonization. *Proc Natl Acad Sci U S A,* **102** (22), 8066–8070.

Gust, A. A., Willmann, R., Desaki, Y., Grabherr, H. M., & Nürnberger, T. (2012). Plant lysm proteins: modules mediating symbiosis and immunity. *Trends Plant Sci.* **17** (8), 495–502.

Gutjahr, C., Novero, M., Guether, M., Montanari, O., Udvardi, M., & Bonfante, P. (2009). Presymbiotic factors released by the arbuscular mycorrhizal fungus gigaspora margarita induce starch accumulation in lotus japonicus roots. *New Phytol.* **183** (1), 53–61.

Gutjahr, C. & Parniske, M. (2013). Cell and developmental biology of arbuscular mycorrhiza symbiosis. *Annu. Rev. Cell Dev. Biol.* **29** (1), 593–617.

Gutjahr, C., Radovanovic, D., Geoffroy, J., Zhang, Q., Siegler, H., Chiapello, M., Casieri, L., An, K., An, G., Guiderdoni, E., Kumar, C. S., Sundaresan, V., Harrison, M. J., & Paszkowski, U. (2012). The half-size abc transporters str1 and str2 are indispensable for mycorrhizal arbuscule formation in rice. *The Plant Journal,* **69** (5), 906–920.

Habib, H. & Fazili, K. M. (2007). Plant protease inhibitors: a defense strategy in plants. *Biotech. Mol. Biol. Rev.* **2** (3), 068–085.

Hagen, F. S., Gray, C. L., O'Hara, P., Grant, F. J., Saari, G. C., Woodbury, R. G., Hart, C. E., Insley, M., Kisiel, W., & Kurachi, K. (1996). Characterization of a cdna coding for human factor vii. *Proc. Natl. Acad. Sci.* **83** (8), 2412–2416.

Harrison, M. J. (2012). Cellular programs for arbuscular mycorrhizal symbiosis. *Curr Opin Plant Biol,* **15** (6), 691–698.

Harrison, M. J., Dewbre, G. R., & Liu, J. (2002). A phosphate transporter from medicago truncatula involved in the acquisition of phosphate released by arbuscular mycorrhizal fungi. *Plant Cell,* **14** (10), 2413–2429.

Haruta, M., Major, I. T., Christopher, M. E., Patton, J. J., & Constabel, C. P. (2001). A kunitz trypsin inhibitor gene family from trembling aspen (populus tremuloides michx.): cloning, functional expression, and induction by wounding and herbivory. *Plant Mol Biol,* **46** (3), 347–359.

Hashimoto, K., Koga, M., Kouhara, H., Arita, N., Hayakawa, T., Kishimoto, T., & Sato, B. (1994). Expression patterns of messenger ribonucleic acids encoding prohormone convertases (pc2 and pc3) in human pituitary adenomas. *Clinical Endocrin.* **41** (2), 185–191.

Hayashi, T., Banba, M., Shimoda, Y., Kouchi, H., Hayashi, M., & Imaizumi-Anraku, H. (2010). A dominant function of ccamk in intracellular accommodation of bacterial and fungal endosymbionts. *Plant J.* **63** (1), 141–154.

He, J., Benedito, V., Wang, M., Murray, J., Zhao, P., Tang, Y., & Udvardi, M. (2009). The medicago truncatula gene expression atlas web server. *BMC Bioinformatics,* **10** (1), 441.

Hedstrom, L. (2002). Serine protease mechanism and specificity. *Chem. Rev.* **102** (12), 4501–4524.

Heidt, S. (2012). Characterisation of an interaction between proteases and kunitz-type protease inhibitors in medicago truncatula. Bachelor Thesis, Karlsruhe Institute of Technology.

Helber, N., Wippel, K., Sauer, N., Schaarschmidt, S., Hause, B., & Requena, N. (2011). A versatile monosaccharide transporter that operates in the arbuscular mycorrhizal fungus glomus sp is crucial for the symbiotic relationship with plants. *Plant Cell,* **23** (10), 3812–3823.

Hemetsberger, C., Herrberger, C., Zechmann, B., Hillmer, M., & Doehlemann, G. (2012). The *Ustilago maydis* effector pep1 suppresses plant immunity by inhibition of host peroxidase activity. *PLoS Pathog,* **8** (5), e1002684.

Hogekamp, C., Arndt, D., Pereira, P. A., Becker, J. D., Hohnjec, N., & Küster, H. (2011). Laser microdissection unravels cell-type-specific transcription in arbuscular mycorrhizal roots, including caat-box transcription factor gene expression correlating with fungal contact and spread. *Plant Physiol.* **157** (4), 2023–2043.

Hohnjec, N., Vieweg, M. F., Pühler, A., Becker, A., & Küster, H. (2005). Overlaps in the transcriptional profiles of medicago truncatula roots inoculated with two different glomus fungi provide insights into the genetic program activated during arbuscular mycorrhiza. *Plant Physiol,* **137** (4), 1283–1301.

Huang, H., Qi, S. D., Qi, F., Wu, C. A., Yang, G. D., & Zheng, C. C. (2010). Ntkti1, a kunitz trypsin inhibitor with antifungal activity from nicotiana tabacum, plays an important role in tobacco's defense response. *FEBS J,* **277** (19), 4076–4088.

Huesgen, P. F. & Overall, C. M. (2012). N- and c-terminal degradomics: new approaches to reveal biological roles for plant proteases from substrate identification. *Physiol Plant,* **145** (1), 5–17.

Innis, M. A., Gelfand, D. H., Sninsky, J. J., & White, T. J., eds **(1990).** *PCR Protocols: a guide to methods and applications.* Academic Press.

Ivanov, S., Fedorova, E. E., Limpens, E., De Mita, S., Genre, A., Bonfante, P., & Bisseling, T. (2012). Rhizobium–legume symbiosis shares an exocytotic pathway required for arbuscule formation. *Proc. Natl. Acad. Sci.* **109** (21), 8316–8321.

Javot, H., Penmetsa, R. V., Terzaghi, N., Cook, D. R., & Harrison, M. J. (2007). A medicago truncatula phosphate transporter indispensable for the arbuscular mycorrhizal symbiosis. *Proc Natl Acad Sci U S A,* **104** (5), 1720–1725.

Jiménez, T., Martin, I., Hernández-Nistal, J., Labrador, E., & Dopico, B. (2008). The accumulation of a kunitz trypsin inhibitor from chickpea (tpi-2) located in cell walls is increased in wounded leaves and elongating epicotyls. *Physiol Plant,* **132** (3), 306–317.

Kamiri, M., Inzé, D., & Depicker, A. (2002). Gateway vectors for agrobacterium-mediated plant transformation. *Trends Plant Sci,* **7** (5), 193–195.

Kamoun, S. (2006). A catalogue of the effector secretome of plant pathogenic oomycetes. *Annu. Rev. Phytopathol.* **44** (1), 41–60. PMID: 16448329.

Kanamori, N., Madsen, L. H., Radutoiu, S., Frantescu, M., Quistgaard, E. M. H., Miwa, H., Downie, J. A., James, E. K., Felle, H. H., Haaning, L. L., Jensen, T. H., Sato, S., Nakamura, Y., Tabata, S., Sandal, N., & Stougaard, J. (2006). A nucleoporin is required for induction of ca2+ spiking in legume nodule development and essential for rhizobial and fungal symbiosis. *Proc. Natl. Acad. Sci.* **103** (2), 359–364.

Kelley, L. A. & Sternberg, M. J. (2009). Protein structure prediction on the web: a case study using the phyre server. *Nat Protoc,* **4** (3), 363–371.

Kistner, C., Winzer, T., Pitzschke, A., Mulder, L., Sato, S., Kaneko, T., Tabata, S., Sandal, N., Stougaard, J., Webb, K. J., Szczyglowski, K., & Parniske, M. (2005). Seven lotus japonicus genes required for transcriptional reprogramming of the root during fungal and bacterial symbiosis. *Plant Cell,* **17** (8), 2217–2229.

Kloppholz, S., Kuhn, H., & Requena, N. (2011). A secreted fungal effector of glomus intraradices promotes symbiotic biotrophy. *Curr Biol,* **21** (14), 1204–1209.

Kobae, Y., Tamura, Y., Takai, S., Banba, M., & Hata, S. (2010). Localized expression of arbuscular mycorrhiza-inducible ammonium transporters in soybean. *Plant Cell Physiol.* **51** (9), 1411–1415.

Koncz, C. & Schell, J. (1986). The promoter of tl-dna gene 5 controls the tissue-specific expression of chimeric genes carried by a novel type of agrobacterium binary vector. *Mol. Gen. Genet,* **204** (3), 383–396.

Kosuta, S., Chabaud, M., Lougnon, G., Gough, C., Dénarié, J., Barker, D. G., & Bécard, G. (2003). A diffusible factor from arbuscular mycorrhizal fungi induces symbiosis-specific mtenod11 expression in roots of medicago truncatula. *Plant Physiol.* **131** (3), 952–962.

Kosuta, S., Hazledine, S., Sun, J., Miwa, H., Morris, R. J., Downie, J. A., & Oldroyd, G. E. D. (2008). Differential and chaotic calcium signatures in the symbiosis signaling pathway of legumes. *Proc. Natl. Acad. Sci.* **105** (28), 9823–9828.

Kretzschmar, T., Kohlen, W., Sasse, J., Borghi, L., Schlegel, M., Bachelier, J. B., Reinhardt, D., Bours, R., Bouwmeester, H. J., & Martinoia, E. (2012). A petunia abc protein controls strigolactone-dependent symbiotic signalling and branching. *Nature,* **483** (7389), 341–344.

Krüger, J., Thomas, C. M., Golstein, C., Dixon, M. S., Smoker, M., Tang, S., Mulder, L., & Jones, J. D. (2002). A tomato cysteine protease required for cf-2-dependent disease resistance and suppression of autonecrosis. *Science,* **296** (5568), 744–747.

Kuhn, H. (2011). *Identification and characterization of Medicago truncatula marker genes for recognition of fungal signals in the arbuscular mycorrhiza symbiosis.* PhD thesis Karlsruhe Institute of Technology (KIT).

Kuhn, H., Küster, H., & Requena, N. (2010). Membrane steroid-binding protein 1 induced by a diffusible fungal signal is critical for mycorrhization in medicago truncatula. *New Phytol,* **185** (3), 716–733.

Kunitz, M. (1946). Crystalline soybean trypsin inhibitor. *J. Gen. Physiol.* **29** (3), 149–154.

Kunitz, M. (1947a). Crystalline soybean trypsin inhibitor: Ii. general properties. *J. Gen. Physiol.* **30** (4), 291–310.

Kunitz, M. (1947b). Isolation of a crystalline protein compound of trypsin and of soybean trypsin-inhibitor. *J Gen Physiol,* **30** (4), 311–320.

Laemmli, U. K. (1970). Cleavage of structural proteins during the assembly of the head of bacteriophage t4. *Nature,* **227** (5259), 680–685.

Larkin, M. A., Blackshields, G., Brown, N. P., Chenna, R., McGettigan, P. A., McWilliam, H., Valentin, F., Wallace, I. M., Wilm, A., Lopez, R.,

Thompson, J. D., Gibson, T. J., & Higgins, D. G. (2007). Clustal w and clustal x version 2.0. *Bioinformatics,* **23** (21), 2947–2948.

Laskowski, M. (1968). The mechansim of formation of a trypsin-trypsin inhibitor complex. *Annu. New York Acad. Scien.* **146** (2), 374–379.

Laskowski, M. & Kato, I. (1980). Protein inhibitors of proteinases. *Annu. Rev. Biochem.* **49** (1), 593–626. PMID: 6996568.

Laskowski, M. & Qasim, M. A. (2000). What can the structures of enzyme-inhibitor complexes tell us about the structures of enzyme substrate complexes? *Biochim Biophys Acta,* **1477** (1–2), 324–337.

Lehfeldt, C., Shirley, A. M., Meyer, K., Ruegger, M. O., Cusumano, J. C., Viitanen, P. V., Strack, D., & Chapple, C. (2000). Cloning of the sng1 gene of arabidopsis reveals a role for a serine carboxypeptidase-like protein as an acyltransferase in secondary metabolism. *Plant Cell,* **12** (8), 1295–1306.

Lerouge, P., Roche, P., Faucher, C., Maillet, F., Truchet, G., Prome, J. C., & Denarie, J. (1990). Symbiotic host-specificity of rhizobium meliloti is determined by a sulphated and acylated glucosamine oligosaccharide signal. *Nature,* **344** (6268), 781–784.

Lesins, K. & Lesins, I. (1979). *Genus Medicago: (Leguminosae) : a taxogenetic study.* W. Junk.

Lévy, J., Bres, C., Geurts, R., Chalhoub, B., Kulikova, O., Duc, G., Journet, E.-P., Ané, J.-M., Lauber, E., Bisseling, T., Dénarié, J., Rosenberg, C., & Debellé, F. (2004). A putative ca2+ and calmodulin-dependent protein kinase required for bacterial and fungal symbioses. *Science,* **303** (5662), 1361–1364.

Li, B. & Fields, S. (1993). Identification of mutations in p53 that affect its binding to sv40 large t antigen by using the yeast two-hybrid system. *The FASEB Journal,* **7** (10), 957–63.

Li, J., Brader, G., & Palva, E. T. (2008). Kunitz trypsin inhibitor: an antagonist of cell death triggered by phytopathogens and fumonisin b1 in arabidopsis. *Mol Plant,* **1** (3), 482–495.

Li, J., Lease, K., Tax, F., & Walker, J. (2001). Brs1, a serine carboxypeptidase, regulates bri1 signaling in arabidopsis thaliana. *Proc. Natl. Acad. Sci.* **98** (10), 5916–5921.

Lievens, S., Goormachtig, S., & Holsters, M. (2004). Nodule-enhanced protease inhibitor gene: emerging patterns of gene expression in nodule development on sesbania rostrata. *J Exp Bot,* **55** (394), 89–97.

Liu, J., Blaylock, L. A., Endre, G., Cho, J., Town, C. D., VandenBosch, K. A., & Harrison, M. J. (2003). Transcript profiling coupled with spatial expression analyses reveals genes involved in distinct developmental stages of an arbuscular mycorrhizal symbiosis. *Plant Cell,* **15** (9), 2106–2123.

Liu, J., Maldonado-Mendoza, I., Lopez-Meyer, M., Cheung, F., Town, C. D., & Harrison, M. J. (2007). Arbuscular mycorrhizal symbiosis is accompanied by local and systemic alterations in gene expression and an increase in disease resistance in the shoots. *Plant J,* **50** (3), 529–544.

Liu, W., Kohlen, W., Lillo, A., Op den Camp, R., Ivanov, S., Hartog, M., Limpens, E., Jamil, M., Smaczniak, C., Kaufmann, K., Yang, W.-C., Hooiveld, G. J., Charnikhova, T., Bouwmeester, H. J., Bisseling, T., & Geurts, R. (2011). Strigolactone biosynthesis in medicago truncatula and rice requires the symbiotic gras-type transcription factors nsp1 and nsp2. *Plant Cell,* **23** (10), 3853–3865.

Logi, C., Sbrana, C., & Giovannetti, M. (1998). Cellular events involved in survival of individual arbuscular mycorrhizal symbionts growing in the absence of the host. *Appl Environ Microbiol,* **64** (9), 3473–349.

Lota, F., Wegmüller, S., Buer, B., Sato, S., Bräutigam, A., Hanf, B., & Bucher, M. (2013). The cis-acting cttc–p1bs module is indicative for gene function of ljvti12, a qb-snare protein gene that is required for arbuscule formation in lotus japonicus. *Plant J.* **74** (2), 280–293.

Lozano-Torres, J. L., Wilbers, R. H., Gawronski, P., Boshoven, J. C., Finkers-Tomczak, A., Cordewener, J. H., America, A. H., Overmars, H. A., Van 't Klooster, J. W., Baranowski, L., Sobczak, M., Ilyas, M., van der Hoorn, R. A., Schots, A., de Wit, P. J., Bakker, J., Goverse, A., & Smant, G. (2012). Dual disease resistance mediated by the immune receptor cf-2 in tomato requires a common virulence target of a fungus and a nematode. *Proc Natl Acad Sci U S A,* **109** (25), 10119–10124.

Maekawa, T., Kusakabe, M., Shimoda, Y., Sato, S., Tabata, S., Murooka, Y., & Hayashi, M. (2008). Polyubiquitin promoter-based binary vectors for overexpression and gene silencing in lotus japonicus. *Mol. Plant Microbe Interact.* **21** (4), 375–382.

Maillet, F., Poinsot, V., Andre, O., Puech-Pages, V., Haouy, A., Gueunier, M., Cromer, L., Giraudet, D., Formey, D., Niebel, A., Martinez, E. A., Driguez, H., Becard, G., & Denarie, J. (2011). Fungal lipochitooligosaccharide symbiotic signals in arbuscular mycorrhiza. *Nature,* **469** (7328), 58–63.

Major, I. T. & Constabel, C. P. (2008). Functional analysis of the kunitz trypsin inhibitor family in poplar reveals biochemical diversity and multiplicity in defense against herbivores. *Plant Physiology,* **146** (3), 888–903.

Manen, J. F., Simon, P., Van Slooten, J. C., Osterås, M., Frutiger, S., & Hughes, G. J. (1991). A nodulin specifically expressed in senescent nodules of winged bean is a protease inhibitor. *Plant Cell,* **3** (3), 259–270.

Martínez, M., Cambra, I., González-Melendi, P., Santamaría, M. E., & Díaz, I. (2012). C1a cysteine-proteases and their inhibitors in plants. *Physiol Plant,* **145** (1), 85–94.

Messinese, E., Mun, J.-H., Yeun, L. H., Jayaraman, D., Rougé, P., Barre, A., Lougnon, G., Schornack, S., Bono, J.-J., Cook, D. R., & Ané, J.-M. (2007). A novel nuclear protein interacts with the symbiotic dmi3 calcium- and calmodulin-dependent protein kinase of medicago truncatula. *Mol. Plant Microbe Interact.* **20** (8), 912–921.

Migliolo, L., de Oliveira, A., Santos, E., Franco, O., & de Sales, M. (2010). Structural and mechanistic insights into a novel non-competitive kunitz trypsin inhibitor from adenanthera pavonina l. seeds with double activity toward serine- and cysteine-proteinases. *J Mol Graph Model,* **29** (2), 148–156.

Miller, R. M., Reinhardt, D. R., & Jastrow, J. D. (1995). External hyphal production of vesicular arbuscular mycorrhizal fungi in pasture and tall grass prairie communities. *Oecologia,* **103**, 17–23.

Minagawa, S., Sugiyama, M., Ishida, M., Nagashima, Y., & Shiomi, K. (2008). Kunitz-type protease inhibitors from acrorhagi of three species of sea anemones. *Comp. Biochem. Physiol.* **150** (2), 240–245.

Mitra, R. M., Gleason, C. A., Edwards, A., Hadfield, J., Downie, J. A., Oldroyd, G. E. D., & Long, S. R. (2004). A ca2+/calmodulin-dependent protein kinase required for symbiotic nodule development: Gene identification by transcript-based cloning. *Proc. Natl. Acad. Sci.* **101** (13), 4701–4705.

Miya, A., Albert, P., Shinya, T., Desaki, Y., Ichimura, K., Shirasu, K., Narusaka, Y., Kawakami, N., Kaku, H., & Shibuya, N. (2007). Cerk1, a lysm receptor kinase, is essential for chitin elicitor signaling in arabidopsis. *Proc. Natl. Acad. Sci.* **104** (49), 19613–19618.

Mueller, A. N., Ziemann, S., Treitschke, S., Aßmann, D., & Doehlemann, G. (2013). Compatibility in the ustilago maydis–maize interaction requires inhibition of host cysteine proteases by the fungal effector pit2 compatibility in the ustilago maydis–maize interaction requires inhibition of host cysteine proteases by the fungal effector pit2. *PLoS Pathog.* **9** (2), e1003177.

Mukherjee, A. & Ané, J.-M. (2010). Germinating spore exudates from arbuscular mycorrhizal fungi: Molecular and developmental responses in plants and their regulation by ethylene. *Mol. Plant Microbe Interact.* **24** (2), 260–270.

Mullen, R. J., Eicher, E. M., & Sidman, R. L. (1976). Purkinje cell degeneration, a new neurological mutation in the mouse. *Proc. Natl. Acad. Sci.* **73** (1), 208–212.

Mundy, J., Svendsen, I., & Hejgaard, J. (1983). Barley α-amylase/subtilisin inhibitor. i. isolation and characterization. *Carlsberg Res. Comm.* **48** (2), 81–90.

Nagahashi, G. & Douds Jr., D. D. (1997). Appressorium formation by am fungi on isolated cell walls of carrot roots. *New Phytol.* **136** (2), 299–304.

Nagahashi, G. & Douds Jr., D. D. (2011). The effects of hydroxy fatty acids on the hyphal branching of germinated spores of {AM} fungi. *Fungal Biology,* **115** (4–5), 351–358.

Navaneetham, D., Wu, W., Li, H., Sinha, D., Tuma, R. F., & Walsh, P. N. (2013). P1 and p2′ site mutations convert protease nexin-2 from a factor xia inhibitor to a plasmin inhibitor. *J. Biochem.* **153** (2), 221–231.

Nielsen, P. K., Bønsager, B. C., Fukuda, K., & Svensson, B. (2004). Barley α-amylase/subtilisin inhibitor: structure, biophysics and protein engineering. *Biochim. Biophys. Acta,* **1696** (2), 157–164.

Nixon, A. & Wood, C. R. (2006). Engineered protein inhibitors of proteases. *Curr. Opin. Drug Discov. Devel.* **9** (2), 261–268.

Nyamsuren, O., Firnhaber, C., Hohnjec, N., Becker, A., Kuester, H., & Krajinski, F. (2007). Suppression of the pathogen-inducible medicago truncatula putative protease-inhibitor mtti2 does not influence root infection by aphanomyces euteiches but results in transcriptional changes from wildtype roots. *Plant Science,* **173** (2), 84–95.

Odeny, D. A., Stich, B., & Gebhardt, C. (2010). Physical organization of mixed protease inhibitor gene clusters, coordinated expression and association with resistance to late blight at the stki locus on potato chromosome iii. *Plant Cell Environ,* **33** (12), 2149–2161.

Oláh, B., Brière, C., Bécard, G., Dénarié, J., & Gough, C. (2005). Nod factors and a diffusible factor from arbuscular mycorrhizal fungi stimulate lateral root formation in medicago truncatula via the dmi1/dmi2 signalling pathway. *Plant J.* **44** (2), 195–207.

Oldroyd, G. E. D. (2013). Speak, friend, and enter: signalling systems that promote beneficial symbiotic associations in plants. *Nat. Rev. Micro.* **11** (4), 252–263.

Oliva, M. L., Ferreira Rda, S., Ferreira, J. G., de Paula, C. A., Salas, C. E., & Sampaio, M. U. (2011). Structural and functional properties of kunitz proteinase inhibitors from leguminosae: a mini review. *Curr Protein Pept Sci,* **12** (5), 348–357.

Oliva, M. L., Silva, M. C., Sallai, R. C., Brito, M. V., & Sampaio, M. U. (2010). A novel subclassification for kunitz proteinase inhibitors from leguminous seeds. *Biochimie,* **92** (11), 1667–1673.

Op den Camp, R., Streng, A., De Mita, S., Cao, Q., Polone, E., Liu, W., Ammiraju, J. S. S., Kudrna, D., Wing, R., Untergasser, A., Bisseling, T., & Geurts, R. (2011). Lysm-type mycorrhizal receptor recruited for rhizobium symbiosis in nonlegume parasponia. *Science,* **331** (6019), 909–912.

Overall, C., Tam, E., Kappelhoff, R., Connor, A., Ewart, T., Morrison, C., Puente, X., Lopez-Otin, C., & Seth, A. (2004). Protease degradomics: mass spectrometry discovery of protease substrates and the clip-chip, a dedicated dna microarray of all human proteases and inhibitors. *Biological Chemistry,* **385**, 493–504.

Ozawa, K. & Laskowski, M., J. (1966). The reactive site of trypsin inhibitors. *J Biol Chem,* **241** (17), 3955–3961.

Penmetsa, V. R., Uribe, P., Anderson, J.and Lichtenzveig, J., Gish, J. C., Nam, Y. W., Engstrom, E., Xu, K., Sckisel, G., Pereira, M., Baek, J. M., Lopez-Meyer, M., Long, S. R., Harrison, M. J., Singh, K. B., Kiss, G. B., & Cook, D. R. (2008). The medicago truncatula ortholog of arabidopsis ein2, sickle, is a negative regulator of symbiotic and pathogenic microbial associations. *Plant J.* **55** (4), 580–595.

Petersen, T. N., Brunak, S., von Heijne, G., & Nielsen, H. (2011). Signalp 4.0: discriminating signal peptides from transmembrane regions. *Nat Methods,* **8** (10), 785–786.

Pettersen, E. F., Goddard, T. D., Huang, C. C., Couch, G. S., Greenblatt, D. M., Meng, E. C., & Ferrin, T. E. (2004). Ucsf chimera–a visualization system for exploratory research and analysis. *J Comput Chem,* **25** (13), 1605–1612.

Piskacek, S., Gregor, M., Nemethova, M., Grabner, M., Kovarik, P., & Piskacek, M. (2007). Nine-amino-acid transactivation domain: Establishment and prediction utilities. *Genomics,* **89** (6), 756 – 768.

Pumplin, N. & Harrison, M. J. (2009). Live-cell imaging reveals periarbuscular membrane domains and organelle location in medicago truncatula roots during arbuscular mycorrhizal symbiosis. *Plant Physiol,* **151** (2), 809–819.

Pumplin, N., Mondo, S. J., Topp, S., Starker, C. G., Gantt, J. S., & Harrison, M. J. (2010). Medicago truncatula vapyrin is a novel protein required for arbuscular mycorrhizal symbiosis. *Plant J,* **61** (3), 482–494.

Pumplin, N., Zhang, X., Noar, R. D., & Harrison, M. J. (2012). Polar localization of a symbiosis-specific phosphate transporter is mediated by a transient reorientation of secretio. *Proc Natl Acad Sci U S A,* **109** (11), 665—672.

Punta, M., Coggill, P. C., Eberhardt, R. Y., Mistry, J., Tate, J., Boursnell, C., Pang, N., Forslund, K., Ceric, G., Clements, J., Heger, A., Holm, L., Sonnhammer, E. L., Eddy, S. R., Bateman, A., & Finn, R. D. (2012). The pfam protein families database. *Nucleic Acids Res,* **40** (Database Issue), D290–D301.

Quandt, H., Pühler, A., & Broer, I. (1993). Transgenic root nodules of vicia hirsuta: a fast and efficient system for the study of gene expression in indeterminate-type nodules. *Mol Plant Microbe Interact,* **6** (6), 699–706.

Rawlings, N. D., Barrett, A. J., & Bateman, A. (2012). Merops: the database of proteolytic enzymes, their substrates and inhibitors. *Nucleic Acids Res,* **40** (D1), 343–350.

Rawlings, N. D., Tolle, D. P., & Barrett, A. J. (2004). Evolutionary families of peptidase inhibitors. *Biochem J,* **378** (Pt 3), 705–716.

Reddy, D. M. R. S., Schorderet, M., Feller, U., & Reinhardt, D. (2007). A petunia mutant affected in intracellular accommodation and morphogenesis of arbuscular mycorrhizal fungi. *Plant J.* **51** (5), 739–750.

Redecker, D., Kodner, R., & Graham, L. E. (2000). Glomalean fungi from the ordovician. *Science,* **289** (5486), 1920–1921.

Remy, W., Taylor, T. N., Hass, H., & Kerp, H. (1994). Four hundred-million-year-old vesicular arbuscular mycorrhizae. *Proc. Natl. Acad. Sci.* **91** (25), 11841–11843.

Requena, N., Serrano, E., Ocón, A., & Breuninger, M. (2007). Plant signals and fungal perception during arbuscular mycorrhiza establishment. *Phytochemistry,* **68** (1), 33–40.

Richau, K. H., Kaschani, F., Verdoes, M., Pansuriya, T. C., Niessen, S., Stuber, K., Colby, T., Overkleeft, H. S., Bogyo, M., & Van der Hoorn, R. A. (2012). Subclassification and biochemical analysis of plant papain-like cysteine proteases displays subfamily-specific characteristics. *Plant Physiol,* **158** (4), 1583–1599.

Rooney, H. C., Van't Klooster, J. W., van der Hoorn, R. A., Joosten, M. H., Jones, J. D., & de Wit, P. J. (2005). Cladosporium avr2 inhibits tomato rcr3 protease required for cf-2-dependent disease resistance. *Science,* **308** (5729), 1783–1786.

Rubinstein, C. V., Gerrienne, P., de la Puente, G. S., Astini, R. A., & Steemans, P. (2010). Early middle ordovician evidence for land plants in argentina (eastern gondwana). *New Phytol.* **188** (2), 365–369.

Ryan, C. A. (1990). Protease inhibitors in plants: Genes for improving defenses against insects and pathogens. *Annu. Rev. Phytopathol.* **28**, 425–449.

Saito, K., Yoshikawa, M., Yano, K., Miwa, H., Uchida, H., Asamizu, E., Sato, S., Tabata, S., Imaizumi-Anraku, H., Umehara, Y., Kouchi, H., Murooka, Y., Szczyglowski, K., Downie, J. A., Parniske, M., Hayashi, M., & Kawaguchi, M. (2007). Nucleoporin85 is required for calcium spiking, fungal and bacterial symbioses, and seed production in lotus japonicus. *Plant Cell,* **19** (2), 610–624.

Salameh, M. A., Soares, A. S., Navaneetham, D., Sinha, D., Walsh, P. N., & Radisky, E. S. (2010). Determinants of affinity and proteolytic stability in interactions of kunitz family protease inhibitors with mesotrypsin. *Journal of Biological Chemistry,* **285** (47), 36884–36896.

Sambrook, J. & Russell, D. W. (2001). *Molecular Cloning – A Laboratory Manual.* Cold Springer Harbor, New York: Cold Springer Harbor Laboratory Press, 3 edition.

Sawers, R. J., Gutjahr, C., & Paszkowski, U. (2007). Cereal mycorrhiza: an ancient symbiosis in modern agriculture. *Trends Plant Sci,* **13** (2), 93–97.

Scannerini, S. & Bonfante-Fasolo, P. (1983). Comparative ultrastructural analysis of mycorrhizal associations. *Canadian Journal of Botany,* **61** (3), 917–943.

Schenck, N. C. & Smith, G. S. (1982). Additional new and unreported species of mycorrhizal fungi additional new and unreported species of mycorrhizal fungi (endogonaceae) from florida. *Mycologia,* **74** (1), 77–92.

Schenkluhn, L., Hohnjec, N., Niehaus, K., Schmitz, U., & Colditz, F. (2010). Differential gel electrophoresis (dige) to quantitatively monitor early symbiosis- and pathogenesis-induced changes of the medicago truncatula root proteome. *J. Proteom.* **73** (4), 753–768.

Schneidman-Duhovny, D., Inbar, Y., Nussinov, R., & Wolfson, H. J. (2005). Patchdock and symmdock: servers for rigid and symmetric docking. *Nucleic Acid. Res.* **33** (W1), 363–367.

Schüßler, A. (2009). Ökologische rolle von pilzen. In: *Rundgespräche der Kommission für Ökologie* volume 37 pp. 97–108. Verlag Dr. Friedrich Pfeil, München.

Schüßler, A., Schwarzott, D., & Walker, C. (2001). A new fungal phylum, the glomeromycota: phylogeny and evolution. *Mycol. Res.* **105** (12), 1413–1421.

Schütze, K., Harter, K., & Chaban, C. (2009). Bimolecular fluorescence complementation (bifc) to study protein–protein interactions in living plant cells. In: *Plant Signal Transduction*, (Pfannschmidt, T., ed) volume 479 of *Methods in Molecular Biology* pp. 189–202. Humana Press.

Schwede, T., Kopp, J., Guex, N., & Peitsch, M. C. (2003). Swiss-model: An automated protein homology-modeling server. *Nucleic Acids Res,* **31** (13), 3381–3385.

Schweitz, H., Bruhn, T., Guillemare, E., Moinier, D., Lancelin, J.-M., Béress, L., & Lazdunski, M. (1995). Kalicludines and kaliseptine. two different classes of sea anemone toxins for voltage sensitive k+ channels. *J. Biol. Chem.* **270** (42), 25121–25126.

Scott, C. J. & Taggart, C. C. (2010). Biologic protease inhibitors as novel therapeutic agents. *Biochimie,* **92** (11), 1681–1688.

Sheokand, S., Dahiya, P., Vincent, J., & Brewin, N. (2005). Modified expression of cysteine protease affects seed germination, vegetative growth and nodule development in transgenic lines of medicago truncatula. *Plant Science,* **169** (5), 966 – 975.

Shindo, T. & van der Hoorn, R. A. L. (2008). Papain-like cysteine proteases: key players at molecular battlefields employed by both plants and their invaders. *Mol. Plant Pathol.* **9** (1), 119–125.

Sieberer, B. J., Chabaud, M., Fournier, J., Timmers, A. C., & Barker, D. G. (2012). A switch in ca2+ spiking signature is concomitant with endosymbiotic microbe entry into cortical root cells of medicago truncatula. *Plant J.* **69** (5), 822–830.

Simon, L., Bousquet, J., Levesque, R. C., & Lalonde, M. (1993). Origin and diversification of endomycorrhizal fungi and coincidence with vascular land plants. *Nature,* **363** (6424), 67–69.

Smith, S. E. & Read, D. J. (2008). *Mycorrhizal Symbiosis.* London: Academic., 3rd edition.

Smith, S. E. & Smith, F. A. (2011). Roles of arbuscular mycorrhizas in plant nutrition and growth: New paradigms from cellular to ecosystem scales. *Annu. Rev. Plant Biol.* **62** (1), 227–250. PMID: 21391813.

Song, H. K. & Suh, S. W. (1998). Kunitz-type soybean trypsin inhibitor revisited: refined structure of its complex with porcine trypsin reveals an insight into the interaction between a homologous inhibitor from erythrina caffra and tissue-type plasminogen activator. *J. Mol. Biol.* **275** (2), 347–363.

Song, J., Win, J., Tian, M., Schornack, S., Kaschani, F., Ilyas, M., van der Hoorn, R. A., & Kamoun, S. (2009). Apoplastic effectors secreted by two unrelated eukaryotic plant pathogens target the tomato defense protease rcr3. *Proc Natl Acad Sci U S A,* **106** (5), 1654–1659.

Srinivasan, A., Giri, A. P., Harsulkar, A. M., Gatehouse, J. A., & Gupta, V. S. (2005). A kunitz trypsin inhibitor from chickpea (cicer arietinum l.) that exerts anti-metabolic effect on podborer (helicoverpa armigera) larvae. *Plant Mol Biol,* **57** (3), 359–374.

St-Arnaud, M., Hamel, C., Vimard, B., Caron, M., & Fortin, J. (1996). Enhanced hyphal growth and spore production of the arbuscular mycorrhizal fungus glomus intraradices in an in vitro system in the absence of host roots. *Mycological Research,* **100** (3), 328 – 332.

Stehle, F., Brandt, W., Stubbs, M. T., Milkowski, C., & Strack, D. (2009). Sinapoyltransferases in the light of molecular evolution. *Phytochemistry,* **70** (15–16), 1652–1662.

Stergiopoulos, I. & de Wit, P. J. (2009). Fungal effector proteins. *Annu. Rev. Phytopathol.* **47** (1), 233–263. PMID: 19400631.

Stracke, S., Kistner, C., Yoshida, S., Mulder, L., Sato, S., Kaneko, T., Tabata, S., Sandal, N., Stougaard, J., Szczyglowski, K., & Parniske, M. (2002). A plant receptor-like kinase required for both bacterial and fungal symbiosis. *Nature,* **417** (6892), 959–962.

Strydom, D. J. (1973). Protease inhibitors as snake venom toxins. *Nat. New. Biol.* **16** (243), 88–89.

Südhof, T. C. & Rothman, J. E. (2009). Membrane fusion: Grappling with snare and sm proteins. *Science,* **323** (5913), 474–477.

Sweet, R. M., Wright, H. T., Janin, J., Chothia, C. H., & Blow, D. M. (1974). Crystal structure of the complex of porcine trypsin with soybean trypsin inhibitor (kunitz) at 2.6-a resolution. *Biochemistry,* **13** (20), 4212–4228.

Takeda, N., Haage, K., Sato, S., Tabata, S., & Parniske, M. (2011). Activation of a lotus japonicus subtilase gene during arbuscular mycorrhiza is dependent on the common symbiosis genes and two cis-active promoter regions. *Molecular Plant-Microbe Interactions,* **24** (6), 662–670.

Takeda, N., Kistner, C., Kosuta, S., Winzer, T., Pitzschke, A., Groth, M., Sato, S., Kaneko, T., Tabata, S., & Parniske, M. (2007). Proteases in plant root symbiosis. *Phytochemistry,* **68** (1), 111–121.

Takeda, N., Maekawa, T., & Hayashi, M. (2012). Nuclear-localized and deregulated calcium- and calmodulin-dependent protein kinase activates rhizobial and mycorrhizal responses in lotus japonicus. *The Plant Cell Online,* **24** (2), 810–822.

Takeda, N., Sato, S., Asamizu, E., Tabata, S., & Parniske, M. (2009). Apoplastic plant subtilases support arbuscular mycorrhiza development in lotus japonicus. *Plant J,* **58** (5), 766–777.

Takeda, N., Tsuzuki, S., Suzaki, T., Parniske, M., & Kawaguchi, M. (2013). Cerberus and nsp1 of lotus japonicus are common symbiosis genes that modulate arbuscular mycorrhiza development. *Plant Cell Physiol.* **54** (10), 1711–1723.

Tamura, K., Peterson, D., Peterson, N., Stecher, G., Nei, M., & Kumar, S. (2011). Mega5: molecular evolutionary genetics analysis using maximum likelihood, evolutionary distance, and maximum parsimony methods. *Mol Biol Evol,* **28** (10), 2731–2739.

Tian, M., Huitema, E., da Cunha, L., Torto-Alalibo, T., & Kamoun, S. (2004). A kazal-like extracellular serine protease inhibitor from phytophthora infestans targets the tomato pathogenesis-related protease p69b. *J. Biol. Chem.* **279** (25), 26370–26377.

Trouvelot, A., Kough, J., & Gianinazzi-Pearson, V. (1986). Mesure du taux de mycorhization va d'un système radiculaire. recherche de méthodes d'estimation ayant une signification fonctionnelle. In: *Physiological and Genetical Aspects of Mycorrhizae*, (Gianinazzi-Pearson, V. & Gianinazzi, S., eds) pp. 217–221. INRA Press Paris.

Turk, B. (2006). Targeting proteases: successes, failures and future prospects. *Nat. Rev. Drug Discov.* **5** (9), 1474–1776.

Untergasser, A. (2008). *RNA Miniprep using CTAB.* Untergasser's Lab.

Valueva, T. A., Revina, T. A., Kladnitskaya, G. V., & Mosolov, V. V. (1998). Kunitz-type proteinase inhibitors from intact and phytophthora-infected potato tubers. *FEBS Lett,* **426** (1), 131–134.

van der Hoorn, R. A. (2008). Plant proteases: from phenotypes to molecular mechanisms. *Annu Rev Plant Biol,* **59**, 191–223.

van der Hoorn, R. A. L., Leeuwenburgh, M. A., Bogyo, M., Joosten, M. H. A. J., & Peck, S. C. (2004). Activity profiling of papain-like cysteine proteases in plants. *Plant Physiol.* **135** (3), 1170–1178.

van der Linde, K., Hemetsberger, C., Kastner, C., Kaschani, F., van der Hoorn, R. A., Kumlehn, J., & Doehlemann, G. (2012a). A maize cystatin suppresses host immunity by inhibiting apoplastic cysteine proteases. *Plant Cell,* **24** (3), 1285–1300.

van der Linde, K., Mueller, A., Hemetsberger, C., Kaschani, F., Van der Hoorn, R. A., & Doehlemann, G. (2012b). The maize cystatin cc9 interacts with apoplastic cysteine proteases. *Plant Signal Behav,* **7** (11), 1397–1401.

Vierheilig, H., Coughlan, A. P., Wyss, U., & Piché, Y. (1998). Ink and vinegar, a simple staining technique for arbuscular-mycorrhizal fungi. *Appl Environ Microbiol,* **64** (12), 5004–5007.

Vincent, J. L. & Brewin, N. J. (2000). Immunolocalization of a cysteine protease in vacuoles, vesicles, and symbiosomes of pea nodule cells. *Plant Physiology,* **123** (2), 521–530.

Voinnet, O., Rivas, S., Mestre, P., & Baulcombe, D. (2003). An enhanced transient expression system in plants based on suppression of gene silencing by the p19 protein of tomato bushy stunt virus. *Plant J.* **33** (5), 949–956.

Vorster, B. J., Schlüter, U., du Plessis, M., van Wyk, S., Makgopa, M. E., Ncube, I., Quain, M. D., Kunert, K., & Foyer, C. H. (2013). The cysteine protease–cysteine protease inhibitor system explored in soybean nodule development. *Agronomy,* **3** (3), 550–570.

Wan, J., Zhang, X.-C., Neece, D., Ramonell, K. M., Clough, S., Kim, S.-y., Stacey, M. G., & Stacey, G. (2008). A lysm receptor-like kinase plays a critical role in chitin signaling and fungal resistance in arabidopsis. *Plant Cell,* **20** (2), 471–481.

Wang, B. & Qiu, Y.-L. (2006). Phylogenetic distribution and evolution of mycorrhizas in land plants. *Mycorrhiza,* **16** (5), 299–363.

Wang, B., Yeun, L. H., Xue, J.-Y., Liu, Y., Ané, J.-M., & Qiu, Y.-L. (2010). Presence of three mycorrhizal genes in the common ancestor of land plants suggests a key role of mycorrhizas in the colonization of land by plants. *New Phytol.* **186** (2), 514–525.

Wang, E., Schornack, S., Marsh, J. F., Gobbato, E., Schwessinger, B., Eastmond, P., Schultze, M., Kamoun, S., & Oldroyd, G. E. D. (2012). A common signaling process that promotes mycorrhizal and oomycete colonization of plants. *Current Biology,* **22** (23), 2242–2246.

Wegel, E., Schauser, L., Sandal, N., Stougaard, J., & Parniske, M. (1998). Mycorrhiza mutants of lotus japonicus define genetically independent steps during symbiotic infection. *Mol. Plant Microbe Interact.* **11** (9), 933–936.

Whipps, J. M. (2004). Prospects and limitations for mycorrhizas in biocontrol of root pathogens. *Can. J. Bot.* **82** (8), 1198–1227.

Wiederstein, M. & Sippl, M. J. (2007). Prosa-web: interactive web service for the recognition of errors in three-dimensional structures of proteins. *Nucleic Acids Res,* **35** (Web Server issue), W407–W410.

Wilson, K. A., Papastoitsis, G., Hartl, P., & Tan-Wilson, A. (1988). Survey of the proteolytic activities degrading the kunitz trypsin inhibitor and glycinin in germinating soybeans. *Plant Physiol.* **88** (2), 355–360.

Wolf, D. H. & Fink, G. R. (1975). Proteinase c (carboxypeptidase y) mutant of yeast. *J. Bacteriol.* **123** (3), 1150–1156.

Wulf, A., Manthey, K., Doll, J., Perlick, A. M., Linke, B., Bekel, T., Meyer, F., Franken, P., Küster, H., & Krajinski, F. (2003). Transcriptional changes in response to arbuscular mycorrhiza development in the model plant medicago truncatula. *Mol Plant Microbe Interact,* **16** (4), 306–314.

Xia, Y. (2004). Proteases in pathogenesis and plant defence. *Cellular Microbiology,* **6** (10), 905–913.

Yamasaki, T., Deguchi, M., Fujimoto, T., Masumura, T., Uno, T., Kanamaru, K., & Yamagata, H. (2006). Rice bifunctional α-amylase/subtilisin inhibitor: Cloning and characterization of the recombinant inhibitor expressed in escherichia coli. *Biosci. Biotechnol. Biochem.* **70** (5), 1200–1209.

Yano, K., Yoshida, S., Müller, J., Singh, S., Banba, M., Vickers, K., Markmann, K., White, C., Schuller, B., Sato, S., Asamizu, E., Tabata, S., Murooka, Y., Perry, J., Wang, T. L., Kawaguchi, M., Imaizumi-Anraku, H., Hayashi, M., & Parniske, M. (2008). Cyclops, a mediator of symbiotic intracellular accommodation. *Proc. Natl. Acad. Sci.* **105** (51), 20540–20545.

Yuan, C.-H., He, Q.-Y., Peng, K., Diao, J.-B., Jiang, L.-P., Tang, X., & Liang, S.-P. (2008). Discovery of a distinct superfamily of kunitz-type toxin (ktt) from tarantulas. *PLoS ONE,* **3** (10), e3414.

Zhang, Q., Blaylock, L. A., & Harrison, M. J. (2010). Two medicago truncatula half-abc transporters are essential for arbuscule development in arbuscular mycorrhizal symbiosis. *Plant Cell,* **22** (5), 1483–1497.

Zhao, R., Dai, H., Qiu, S., Li, T., He, Y., Ma, Y., Chen, Z., Wu, Y., Li, W., & Cao, Z. (2011). Sdpi, the first functionally characterized kunitz-type trypsin inhibitor from scorpion venom. *PLoS ONE,* **6** (11), e27548.

Zhou, A. & Li, J. (2005). Arabidopsis brs1 is a secreted and active serine carboxypeptidase. *J. Biol. Chem.* **280** (42), 35554–35561.

Župunski, V., Kordiš, D., & Gubenšek, F. (2003). Adaptive evolution in the snake venom kunitz/bpti protein family. *FEBS Letters,* **547** (1–3), 131–136.

APPENDIX

Supplementary Figure S1

a

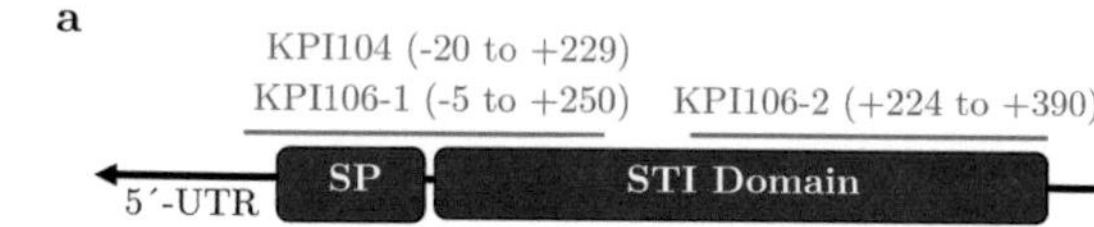

b

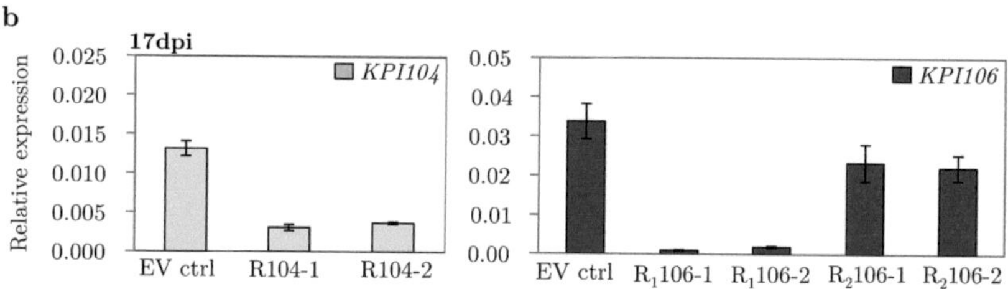

c

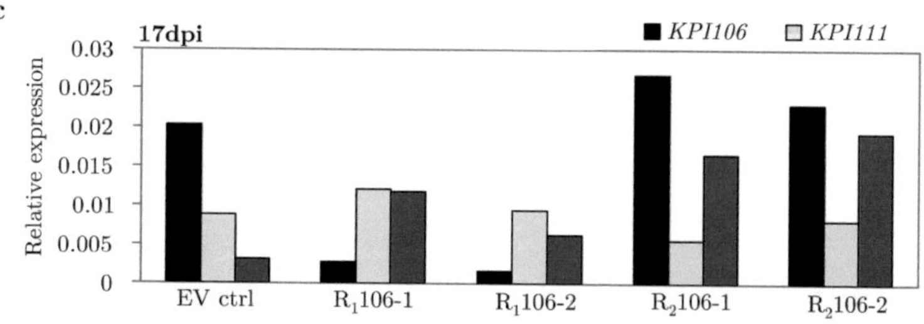

Figure S1: RNAi-mediated silencing of KPI104 and KPI106. **(a)** Localization of the RNAi-targets used to silence *KPI104* and *KPI106*. Numbers indicate basepairs respective to the ATG. **(b)** Real Time PCR of mycorrhizal RNAi roots (17 dpi) revealed successful silencing of each inhibitor. RNAi-targets localized within the 5´-UTR showed a more effective silencing than targeting within the conserved STI domain of KPI106. **(c)** Relative expression of homologous KPIs in R-KPI106 lines at 17 dpi revealed no cross-silencing of KPI104 (light gray) and KPI111(dark gray) in RKPI106 lines, whereas KPI106 was significantly downregulated. Columns represent one biological replicate.

Supplementary Figure S2

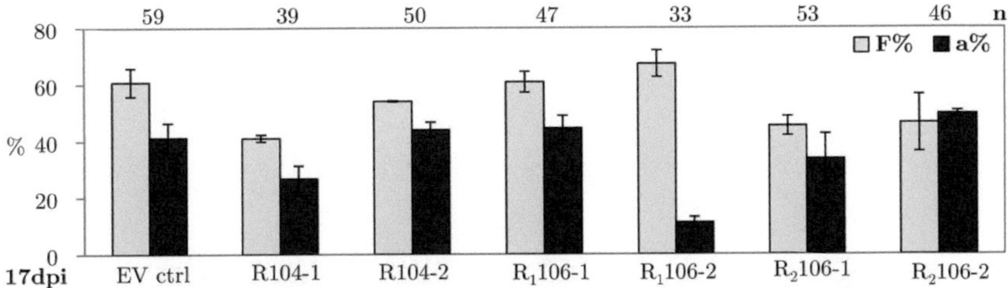

Figure S2: RNAi-mediated silencing of KPI104 and KPI106 revealed no mycorrhizal phenotype. Quantification of mycorrhizal colonization at 17 dpi of KPI104 and KPI106 RNAi hairy root lines revealed no significant alterations in arbuscule abundance when compared to the empty vector control. F%: Frequency of mycorrhizal colonization, a%: arbuscule abundance, n: number of root fragments.

Supplementary Figure S3

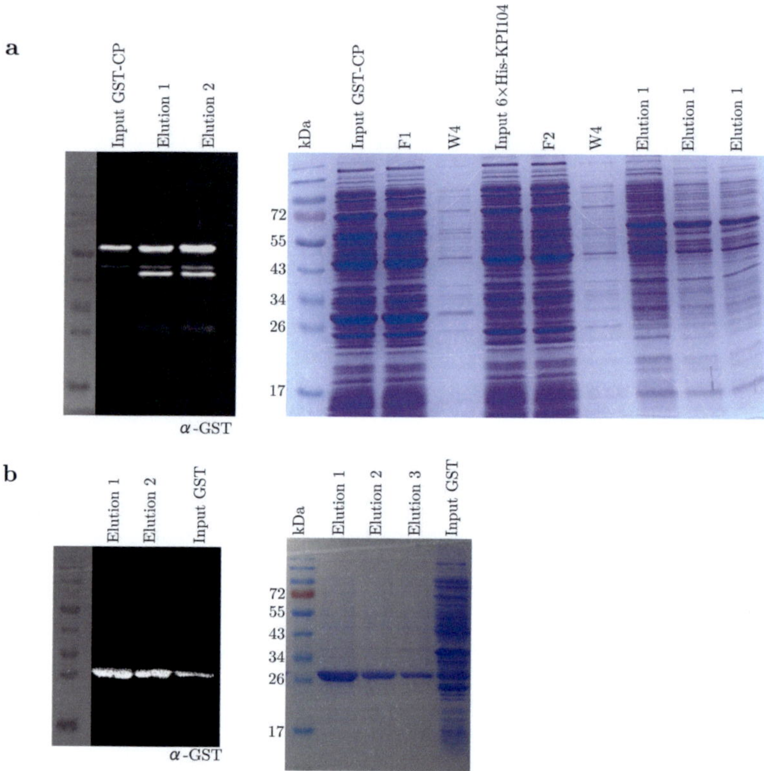

Figure S3: GST-CP pulldown assay. **(a)** Western blot detection of GST-CP (63 kDa) probed with α-GST. The first lane shows the soluble fraction derived from a crude extract of the *E. coli* strain expressing GST-CP that was used as input. Lane 2 and 3 represent the first two eluting fractions of GST-CP after purification by resin-bound glutathione. The occurrence of several distinct bands in those lanes might represent internal cleavage processes. Right: The Coomassie stained gel was loaded with samples of the different steps of the GST-CP pulldown assay. F: Flow-through; W: washing step. The detailed description of the procedure can be reviewed in section 5.5.5. **(b)** Free GST (27 kDa) was used as bait for a control pulldown. The α-GST western blot shows the input fraction, as well as the first two eluting fractions. Right: The according fractions within the Coomassie stained gel.

Supplementary Figure S4

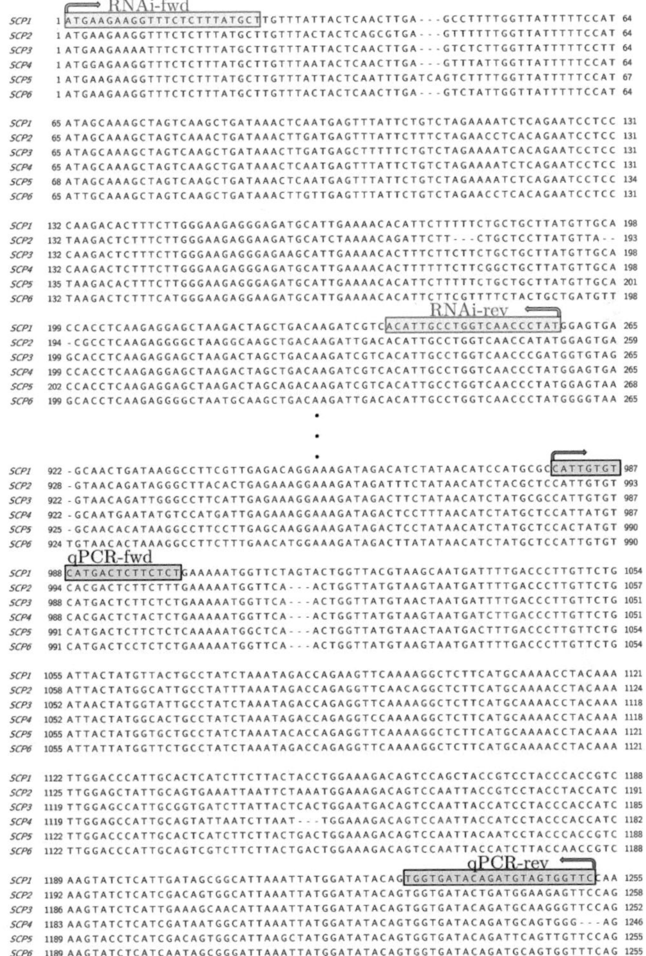

Figure S4: Alignment of cDNA sequences of homologous SCP genes of *M. truncatula*. Highlighted are the primer sequences used for amplification of the RNAi-target (RNAi-fwd; RNAi-rev), as well as the primer pair used for quantification of the *SCP* transcripts by Real Time PCR (qPCR-fwd; qPCR-rev).

Supplementary Figure S5

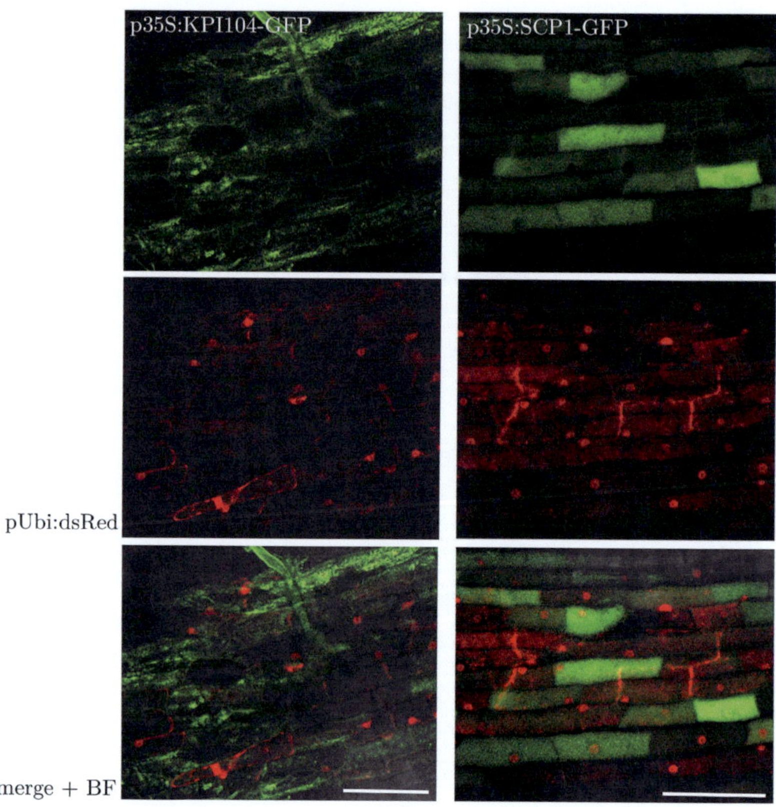

Figure S5: Expression of GFP fusion proteins in *M. truncatula* hairy roots. Shown are confocal images of p35S:KPI104-GFP and p35:SCP1-GFP hairy roots, 17 dpi colonized with *R. irregularis*. pUbi:DsRed served as localization control to label the plant nucleus and cytoplasm. No distinct GFP signal was observed, in contrast to the high endogenous autofluorescence of the hairy roots.

Supplementary Table S1

Table S1: Yeast two-hybrid analysis of KPI106bait. The list contains the KPI106bait interacting clones with the highest color intensities in the X-α-Gal test that were chosen for further investigation. Inserts of the cDNA-library clones were analyzed for in frame position with respect to the ATG of the prey vector pGADT7-Rec and sequences was blasted against *M. truncatula* Gene Index. Clones 69 and 120 encode partial C-terminal fragments of CP.

Clone #	Putative annotation	TC annotation	Score identity
38	reversibly glycosylated protein	medicago|TC121226	99%
55	Chromosome 5, scaffold_2, whole genome shotgun sequence	medicago|TC191584	99%
68	Peptidylprolyl isomerase, FKBP-type	medicago|TC116547	100%
69, 120	Cysteine protease	medicago|TC133093	99%
76	Formyl transferase, N-terminal	medicago|TC113411	98%
6, 80, 125,	Elongation factor 1 alpha	medicago|TC114384	99%
96	Expressed protein Arabidopsis thaliana	medicago|TC189860	100%
124	Ubiquitin	medicago|TC114544	72%
150	Expressed protein Arabidopsis thaliana	medicago|TC196134	99%
151	Probable protein disulfide-isomerase A6 precursor	medicago|TC116161	96%
156	Alpha-N-acetylglucosaminidase	medicago|TC129087	99%
2	Proline rich protein	medicago|TC188749	100%
11	Remorin	medicago|EY476177	59%
19, 60	Disulfide Isomerase	medicago|TC175649	97%
30	Nodule specific cysteine rich peptide	medicago|TC186172	76%

Curriculum Vitae

Persönliches

Vorname	Stefanie Sarah
Nachname	Hirsch, geb. Rech

Bildungsweg

06/2010–12/2013	**Promotion**, Karlsruher Institut für Technologie (KIT), Botanisches Institut
01/2010–05/2010	Wissenschaftliche Mitarbeiterin KIT, Botanisches Institut
03/2009–12/2009	**Diplomarbeit**, KIT, Botanisches Institut, Titel: "Elucidating the function of genes encoding secreted proteins induced during early establishment of arbuscular mycorrhizal symbiosis"
10/2004–12/2009	**Biologiestudium**, Universität Karlsruhe (TH) Abschluss: Diplom-Biologin Note: 1,0 mit Auszeichnung Schwerpunkte: Mikrobiologie, Biochemie, Zoologie, Ingenieurbiologie
10/2003–09/2004	Zahnmedizinstudium (2 Semester), Universität Freiburg
1994–2003	**Abitur**, Fürstenberg-Gymnasium, Donaueschingen

Publikation

Rech, S. S., Heidt, S., Requena, N. (2013). A tandem Kunitz protease inhibitor (KPI106)–serine carboxypeptidase (SCP1) controls mycorrhiza establishment and arbuscule development in *Medicago truncatula*. *Plant J.*, **75**(5), 711–725.

Auszeichung

01/2011–12/2012	Promotionsstipendium der Landesgraduiertenförderung Baden-Württemberg

Konferenzen

09/2010	Poster: "Functional characterization of a secreted protease inhibitor induced during early establishment of arbuscular mycorrhizal symbiosis", 2nd PhD Symposium KIT, Karlsruhe
04/2011	Poster: "Kunitz type protease inhibitors regulate arbuscule development in mycorrhizal symbiosis", Jahrestagung der Vereinigung für Allgemeine und Angewandte Mikrobiologie (VAAM) in Karlsruhe
09/2011	Poster: "Kunitz-type protease inhibitors are involved in arbuscule development in mycorrhizal symbiosis", Internationale Botanikertagung, Berlin
09/2012	Vortrag: "Kunitz-type protease inhibitors are involved in arbuscule development in mycorrhizal symbiosis", 1st International Mycorrhiza Meeting, München